Tomatoes in the Tropics

IADS DEVELOPMENT-ORIENTED
LITERATURE SERIES
Steven A. Breth, series editor

Tomatoes in the Tropics,
Ruben L. Villareal

*Successful Seed Programs: A Planning and Management
Guide,* compiled and edited by Johnson E. Douglas

*Small Farm Development: Understanding and Improving
Farming Systems in the Humid Tropics,* Richard R. Harwood

*Rice in the Tropics: A Guide to the Development
of National Programs,* Robert F. Chandler, Jr.

Tomatoes in the Tropics
Ruben L. Villareal

Routledge
Taylor & Francis Group

LONDON AND NEW YORK

First published 1980 by Westview Press

Published 2019 by Routledge
52 Vanderbilt Avenue, New York, NY 10017
2 Park Square, Milton Park, Abingdon, Oxon OX14 4RN

Routledge is an imprint of the Taylor & Francis Group, an informa business

Copyright © 1980 by the International Agricultural Development Service

Library of Congess Cataloging in Publications Data
Villareal, Ruben L
    Tomatoes in the Tropics.
    (IADS development-oriented literature series)
    Bibliography: p.
    1. Tomatoes–Tropics. I. Title. II. Series: United States. International Agricultural Development Service. IADS development-oriented literature series.
SB349.V54   338.1'756420913   80-16446

ISBN 13: 978-0-367-27406-1 (hbk)
ISBN 13: 978-0-367-27444-3 (pbk)

Contents

Part 1
Administrative Considerations

Appendixes

Figures

Tables

Foreword

Recent advances in the technology of tomato production offer exciting possibilities for substantially increasing the use of this important vegetable in the tropics. Several serious bottlenecks that have kept the commodity from being grown widely and in quantity have been removed or considerably diminished. Government decision makers, produce-market operators, and farmers should be aware that the tomato can be expected to become increasingly plentiful if actions are taken to foster production.

The tomato is a nearly universally popular vegetable. Wherever it can be grown in quantity and at a favorable price it raises the quality of diets. It is a versatile commodity that can be eaten fresh or processed for use in a wide array of products. The number of ways it can be used to improve the flavor and character of other foods is seemingly endless.

Production of the tomato can be an especially profitable way to utilize limited land resources and abundant labor. It can be grown in a household garden, it can contribute substantially to the family's income on an extremely small area of land, or it can be grown on a large scale for urban markets and for processing. The tomato fits well into many cropping patterns and may bring in needed cash during periods when cereals and other staples cannot be grown and when labor is surplus. Near large towns and cities the fresh vegetable market can be particularly lucrative.

Viewed in a larger context the results that have been obtained with the tomato demonstrate the high payoff that can be obtained from properly oriented agricultural research. A commodity need not be grown on millions of hectares to justify research attention. Often the degree of research success can be attributed to the keen insights of scientists and the sharpness of their focus on limiting factors rather than on the total size of their budgets.

Ruben Villareal, the author of this fourth book in the IADS Development-Oriented Literature Series, is an authority on horticultural crops in the tropics. He has played a leading role in launching national research and extension programs in vegetable crops in the Philippines. At the Asian Vegetable Research and Development Center in Taiwan, which he joined in 1972, he is responsible for breeding research on tomatoes and sweet potatoes. The support of the Rockefeller Foundation in the development of this book is gratefully acknowledged.

A. Colin McClung
President
International Agricultural Development Service

Preface

Tomatoes, which are grown both in home gardens and commercially, are one of the world's most popular vegetables. They are good sources of vitamins A and C and can help alleviate deficiencies of these vitamins in many developing countries. Because tomato growing is labor intensive, tomatoes are an attractive cash crop for small farmers and a potential source of rural employment.

In most developing countries, however, cereals have received the highest research and production priorities. The relatively limited emphasis on other crops, such as tomatoes, is understandable as the first need of these countries is staple foods to meet the demands created by rapidly increasing populations. But as staple food production accelerates many countries may find that tomato growing has distinct economic and nutritional advantages.

This book attempts to put together in a single volume useful information about this crop in the tropics. Policymakers and administrators who do not have time to assemble and read numerous reports are the principal target audience, but the book should be equally useful as an introduction to scientists who may implement a tomato program. Many scientists have difficulty finding information on tomatoes because the literature is not readily available in tropical countries. Furthermore, literature related to tomato growing is scattered among unpublished and published theses, abstracts, bulletins, journals, annual reviews, and some limited editions of books. This situation retards the extension of new information to tomato farmers and backyard gardeners.

The three chapters of Part 1 are intended primarily for policy-makers and administrators. Chapter 1 discusses the value and potential of tomatoes in the tropics. Chapter 2 presents in some detail the successes of three national programs and how others can benefit from these experiences. Chapter 3 is intended to assist in an assessment of needs and opportunities for expanded production.

Although Part 2 is mainly for scientists, decision makers and administrators also may find it of value. Chapter 4 outlines the origin, distribution, and uses of tomatoes. Chapters 5 and 6 describe varieties, seed production and distribution, and cultural and pest management practices. Chapter 7 treats postharvest technology and marketing. Finally, Chapter 8 identifies some important research topics that deserve attention. The appendixes provide sources of assistance and a list of tomato scientists. There is also a glossary of terms and an annotated bibliography.

In preparation for writing this book, I visited major tomato production areas and experiment stations in tropical Africa and Latin America. These trips complemented my familiarity with tomato-growing practices in tropical Asia and the Pacific Islands. I am grateful to the many tomato scientists and farmers in the tropics for their unselfish assistance in providing information.

The Asian Vegetable Research and Development Center (AVRDC) generously supplied several illustrations and both published and unpublished information. AVRDC's first two directors (R. F. Chandler, Jr., and J. C. Moomaw) and staff provided advice, support, and encouragement. AVRDC supported the author's sabbatical leave at Cornell University that made this project possible. Cornell University, with its excellent library facilities and a staff with tropical experience, provided an ideal location.

The manuscript was reviewed by G. Hernandez Bravo, national vegetable coordinator at the Instituto Nacional de Investigaciones Agrícolas, Mexico; R. W. Richardson, former director of agricultural sciences for the Rockefeller Foundation; H. M. Munger, professor of vegetable crops and plant breeding at Cornell University; and G. F. Wilson, agronomist at the International Institute of Tropical Agriculture. In addition, S. A. Breth, program officer, IADS, and W. C. Kelly, professor of vegetable crops at Cornell University, gave helpful guidance while the manuscript was being written. I am thankful to all of them. But I assume full responsibility for any errors or omissions.

To Professor William B. Ward of the Department of Communication Arts at Cornell University, I owe my thanks and gratitude for expertly and patiently editing the final version of the manuscript.

I am especially grateful to my family and particularly my wife, Corazon PeBenito Villareal, for patient understanding, constant encouragement, and never-ending confidence.

Ruben L. Villareal
Asian Vegetable Research and Development Center

Part 1

Administrative Considerations

1
Potential for the Tropics

Tomatoes are the world's most widely grown vegetable, other than the white potato. Commercially, 45 million metric tons of tomatoes are produced each year from 2.2 million hectares, but only 15 percent of the output occurs in the tropics. (These figures exclude the large amount of tomatoes grown in home gardens.) Various groups of scientists who have attempted to set priorities for vegetable crops in tropical countries consistently have ranked tomatoes first for increased production and more intensive research.

The potential for tomatoes in the tropics is great. More widespread cultivation of tomatoes could generate rural employment, stimulate urban employment, expand exports, improve nutrition of the people, and increase the income of farmers.

Generate rural employment

Tomato production requires two to three times or more labor per hectare than rice, the principal food crop in heavily populated tropical countries (Table 1). In Taiwan for example, production of processing tomatoes uses over 2000 hours of labor per hectare and production of fresh-market tomatoes uses 8000 hours compared with less than 800 hours for one hectare of rice. Taiwan, in the mid-1970s, needed over 60,000 months of labor annually for the production of tomatoes. In Mexico, fresh-market production in Culiacan, Sinaloa, employs thousands of workers in activities ranging from seedling production in greenhouses to packaging tomatoes in packing houses. The

Table 1
Hours of labor required per hectare for producing tomatoes and rice in nine countries

	Tomatoes	Rice
Colombia	4000	280
India	3384	632-736
Indonesia	3000-6000	1920
Japan	7040	371
Korea	2240-3200	1008-1112
Nigeria	3700-5700	1200-3600
Philippines	1384-2160	688-880
Taiwan	8020	761
Thailand	799-1844	648-936

Note: The data are from scientists in each country except the rice data for India, Indonesia, Korea, Philippines, Taiwan, and Thailand, which were supplied by the International Rice Research Institute.

situation is similar in other tropical countries such as Brazil, Colombia, India, and the Philippines where sizable amounts of tomatoes are grown. Tomato production has a great potential for using underemployed rural workers and increasing their income.

Stimulate urban employment

Production of fresh-market or processing tomatoes provides business opportunities for manufacturers of fertilizers, pesticides, sprayers, and containers (tin cans and crates of bamboo, cardboard, plastic, or wood); for producers of bamboo or wooden poles for propping or staking; and for dealers in seeds and farm implements. The growth of these enterprises will expand urban employment and rural employment as well, if factories are dispersed in the countryside. The business aspect of the industry requires additional people—from clerks to executives—to handle advertising, marketing, and day-to-day activities in offices and factories. Even small-scale production stimulates urban employment as a swelling volume of tomatoes reaches urban markets.

In Mexico tomato production employs thousands of workers in various aspects of the industry from planting seedlings to marketing and distribution. (Source: CIAPAN.)

Tomato production can stimulate related enterprises such as the manufacture of wooden crates (*top*, near Bogotá, Colombia), and the production of props or stakes (*bottom*, near Culiacan, Mexico). (Sources: Instituto Colombiano Agropecuario and CIAPAN.)

Expand exports

A number of tropical countries export fresh or processed tomatoes to developed countries. An overview of international trade in fresh-market tomatoes is presented in Table 2.

Tomatoes have become increasingly expensive in developed countries. In 1978, a ton of processing tomatoes cost about US$160 in Japan and at least $54 in the United States. But in Taiwan, for example, the cost was only about $26 a ton. Because of such cost differentials, Taiwan's export of processed tomatoes increased one-hundred-fold between 1969 and 1978 to about $20 million a year. Inflation and the rising cost of labor will probably make tomatoes even more costly in developed countries in the future.

Imports of fresh-market tomatoes are rising in developed countries. The United States and Canada import tomatoes from

Table 2
International trade in fresh-market tomatoes, 1965 and average annual 1973–1977 (million U.S. dollars)

	Imports		Exports	
	1965	1973–1977	1965	1973–1977
World	268.4	724.6	225.5	571.2
Developed	245.7	644.7	131.3	324.2
North America	46.3	137.9	10.0	279.4
Western Europe	199.4	506.8	121.0	294.6
Other	—	—	0.3	1.6
Developing	7.4	16.3	68.1	187.2
Africa	1.7	0.5	28.2	38.8
Latin America	1.0	2.1	35.9	138.2
Near East	3.2	9.9	3.4	8.7
Far East	1.3	3.3	0.4	1.4
Other	0.1	0.5	0.2	0.05
Centrally planned	15.3	63.7	26.0	59.8
Asian	—	—	—	2.3
Europe, USSR	15.3	63.7	26.0	57.9

Source: FAO, 1966–1978. *FAO Trade Yearbook, 1965–1977*, Rome.

Mexico during the winter season. Western European countries import from Egypt, Morocco, Jordan, and such European countries as the Netherlands and Spain. Hong Kong and Singapore buy from Indonesia, Malaysia, and Taiwan; and trial shipments to Hong Kong have been made from the Philippines. Japan imports processed tomatoes from Taiwan and is a potential market for fresh-market tomatoes from other Asian nations.

Export arrangements usually start with an agreement between a foreign company and a local company to produce tomatoes or manufacture tomato products. When the operation succeeds, the local company buys out the shares of the foreign company. As this happens, other local companies establish tomato production or manufacturing enterprises. Developing countries that have good growing conditions for tomatoes could increase their foreign exchange earnings through exports.

Increase net returns to farmers

Like most vegetable crops, tomatoes can give farmers high income per hectare, especially if the harvests are marketed efficiently. In Taiwan, fresh-market tomatoes, whether grown during the summer or winter, are more profitable than rice—the major competing crop (Figure 1). Although winter processing tomatoes are less profitable than rice, many Taiwanese farmers grow a tomato crop in the winter because weather conditions are less risky than in the summer and because processing tomatoes have an assured market.

Improved nutrition

Although the potential for tomatoes in the tropics lies primarily in increased income and employment, tomatoes can contribute to better nutrition. A publication of the League for International Food Education estimates that tomatoes supply almost as many calories per hectare as rice and more protein. In a study in Hawaii, when the nutrient yields of 16 crops proposed for inclusion in a 42-square-meter model kitchen garden were expressed on a square-meter-per-day basis, tomatoes

US dollars

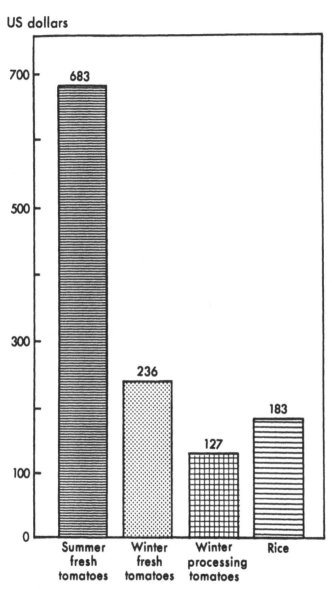

Figure 1. Net returns per hectare for summer tomatoes, winter tomatoes, and rice in Taiwan, 1976–1977. (Source: AVRDC.)

ranked eighth highest in protein and vitamin C, ninth in iron, and thirteenth in vitamin A.

The concentration of beta carotene, a precursor of vitamin A, can be increased at least tenfold in tomatoes through breeding. This trait is governed by a single dominant gene that can be easily transferred to local varieties in the tropics. Unfortunately, tomatoes that have a high concentration of beta carotene are orange-red instead of the familiar tomato-red. In the United States such a color change would reduce tomato acceptability.

Vitamin C content can be increased at least five times, but high vitamin C content in tomatoes has been associated with low yield and small or poorly shaped fruit. If, however, varieties could be bred with a higher content of vitamins A and C combined with other desirable attributes of a tropical tomato and be acceptable to consumers in the tropics, such varieties might have a tremendous impact in alleviating vitamin A and C deficiencies in developing countries.

Tomatoes can be a major source of vitamins and minerals if consumed in large quantities, as they are in the United States. Greater consumption of tomatoes creates a larger demand, thereby encouraging more local production.

Versatility

Few agricultural products lend themselves to as many uses as tomatoes. They are served raw, baked, stewed, fried, pickled, as a sauce, or in combination with other foods. They can be used as an ingredient in both the kitchen and the food manufacturing industry, and they can be processed whole or as a paste, sauce, juice, or powder.

Suitable for home gardens

Like many vegetable crops, tomatoes can be grown efficiently both on a commercial scale and in home gardens. Home gardeners escape marketing problems—the most frustrating aspect of commercial growing. For home gardeners spoilage is not consequential because the tomatoes are harvested and eaten as they ripen. Financing is unimportant in home gardening, a major difference from commercial growing, which requires large

amounts of money to start operations. In addition, home gardeners plant tomatoes as only one of several vegetables that the
family likes to eat. Cherry tomatoes, large-fruited tomatoes, or
even a few rows of processing tomatoes for home canning can
be grown. Commercial growers, on the other hand, must investi-
gate the demand for the type of tomatoes they wish to grow.

Barriers to tomato growing

Inadequate marketing systems, wide seasonal fluctuations in
supplies and prices, low yields, and postharvest losses are problems for tomato growers in the tropics and major reasons they
are often reluctant to increase plantings.

Inadequate marketing systems. In countries where markets
and distribution systems are rudimentary, small farmers often
find themselves at the mercy of middlemen who can dictate
prices. Farmers who bring their baskets of tomatoes to the market and need cash badly may sell their produce at a loss rather
than take it back home. This situation could be alleviated by
markets that function more effectively.

Seasonal fluctuations in supplies and prices. Retail prices in
the tropics soar when the supply of tomatoes is scarce and fall
rapidly when the supply is plentiful (Table 3). The period of
low supply is usually the rainy season when tomatoes are more
difficult to produce. Extreme seasonal price variations make
peak-season production unattractive, especially if yields are low.
On the other hand, off-season production is highly risky. Thus,
unless demand grows, assuring better returns during the peak
season, only large landholders and risk-takers are likely to be
willing to expand tomato production.

Postharvest losses. Improper handling of tomatoes after harvest lowers quality and commonly causes losses ranging from
30 to 45 percent. This waste, coupled with unpredictable prices,
accentuates variability in farm income, which discourages small
farmers from growing tomatoes for market. (Postharvest handling is discussed in Chapter 7.)

Low yields. Like most vegetable crops in the tropics, tomatoes are normally produced in the mountains or during the cool
season in the lowlands. Unadapted varieties are a serious problem. Seeds of imported varieties bred for temperate climates

Table 3

Annual range of retail prices for tomatoes in several cities in the tropics (1978 data)

	Price/kg		Ratio
	Local currency	U.S. $	high/low
Coimbatore, India	0.45-4 rupees	0.06-0.50	8:1
Bangkok, Thailand	0.5-18 bahts	0.025-0.90	36:1
Manila, Philippines	0.5-5 pesos	0.07-0.70	10:1
Dakar, Senegal	60-500 francs	0.30-2.56	8:1
Accra, Ghana	0.5-3 cedis	0.19-1.19	6:1
Ibadan, Nigeria	0.1-2 naira	0.16-3.10	20:1
Culiacan, Mexico	2.5-25 pesos	0.12-1.22	10:1
San Jose, Costa Rica	6.6-19.8 colones	0.78-2.33	3:1
Sao Paulo, Brazil	6-25 cruzeiros	0.30-1.23	4:1

Source: Data collected by G. F. Wilson (for Nigeria) and R. L. Villareal (other countries).

are often the only ones available to farmers, but yields from these varieties are often low and erratic, especially during the hot season. In addition, the lack of appropriate practices and pest management techniques for the wet season and dry season is a barrier to successful production in the tropics. Low yields increase the risk of growing tomatoes. Small farmers will hesitate to plant larger areas to tomatoes unless research leads to higher and more dependable yields.

Lack of vegetable research. The investment of developing countries in agricultural research, education, and production services is low by any measure—percentage of gross national product, per capita per year, or value of agricultural production. To make matters worse, vegetables in these countries rank low in food production priorities compared with staple crops such as rice, maize, and wheat. Consequently, vegetable yields have risen more slowly than yields of cereals (Table 4). Throughout the tropics, the vegetable industry suffers because too few trained researchers are conducting experiments and too few extension personnel are bringing information to farmers and gardeners. In most developing countries cereals are the primary

Table 4

Changes in yields of important cereals and vegetable crops in developing countries, 1961–1965 and 1975

| | Yield | | Annual |
	1961–1965 t/ha	1975 t/ha	increase (%)
Rice	1.61	1.98	2.3
Maize	1.13	1.32	1.7
Wheat	0.98	1.30	3.2
Cabbage	10.40	10.91	0.5
Green beans	3.42	3.86	1.3
Sweet potatoes	7.09	7.41	0.4
Tomatoes	11.63	13.17	1.3

Source: FAO, 1977, *FAO Production Yearbook 1976,* Rome.

responsibility of trained personnel. If they are assigned to work on vegetables, they must deal with so many kinds that they have difficulty developing expertise.

Attempts at solutions

Good organization and planning can remedy many problems associated with commercial tomato growing. In some tropical countries tomato growing is a flourishing industry (see Chapter 2).

There are several ways tomato farmers can be helped to achieve higher income and lower risk. Governments and private industry can provide contract or guaranteed prices, input subsidies, easier credit, better transport, and improved markets. They can also finance training programs for researchers and managers. With adequate support, agricultural research can raise the yield potential of tomatoes through the creation of improved varieties and development of effective and economical management practices. When farmers' risks decrease and profits improve, they will be encouraged to plant larger areas of tomatoes.

The steps leading to commercial tomato growing vary from country to country. In Taiwan and Mexico, for instance, a processing and fresh-market tomato industry developed almost independent of government influence. These industries prospered principally as a result of private initiative. In most developing countries, however, government prodding and assistance is essential to initiate a food production program.

The success of the "green revolution" in raising output of cereals has encouraged many governments in developing countries to devote part of their resources to other crops. In addition, private food industries in developed countries have seen the potential of growing tomatoes for domestic or export markets. Consequently, they have organized agricultural companies in tropical countries with combined local and foreign investment.

2
Successful Programs

Despite the difficulties associated with growing tomatoes in the tropics, tomato output increased in a number of tropical countries during the 1970s. The yields are low by temperate standards, but impressive for tropical conditions. Although there are no nationwide tomato production programs, successful regional and local programs exist. Examples from the Philippines, Taiwan, and Mexico illustrate that when the right ingredients are available tomato production on a small or large scale can be successful in the tropics.

In the southern Philippines, a government program involves about 700 farmers who operate farms of 1 to 2 hectares. They produce fresh-market tomatoes that are transported 750 kilometers to market by ships and airplanes. In Taiwan, private industry took the initiative in importing a successful model for producing and processing tomatoes from Japan. In Mexico, private enterprises produce fresh-market tomatoes on a large scale (500 to 1000 hectares per operation) for export to the United States and Canada.

In both the Philippines and Taiwan small farmers are the beneficiaries. The patterns established in these countries could be duplicated readily in other tropical areas where farm size is generally small. Transferring the Mexican model to other areas may be more difficult because of the huge amount of capital needed for production and marketing on such a large scale. Nevertheless, the Mexican experience shows what can be done with the right climate, technology, market, and capital.

PHILIPPINES

In the Philippines heavy tomato production from March to May causes a market glut and low prices (Figure 2); during the rest of the year supplies are limited and prices are high. The average yield has been low (Figure 3).

Seasonality of supply and low yields are true for other vegetables, too. The major causes of low productivity have been shortage of improved seeds, poor growing practices, and a lack of well-trained and adequately supported extension workers. Vegetable yields started to rise after 1968 when the government began paying attention to research on and production of crops other than rice.

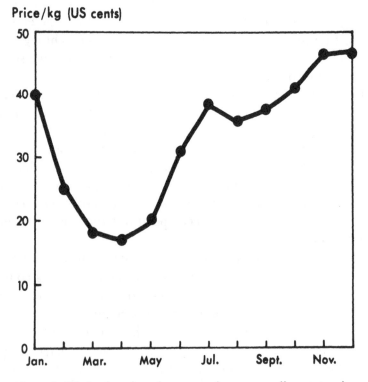

Figure 2. Wholesale price of tomatoes in seven trading centers in the Philippines, 1977. (Source: Bureau of Agricultural Economics, Philippines.)

Yield (t/ha)

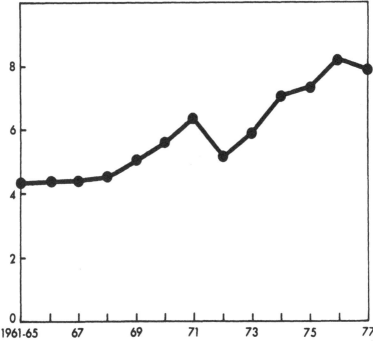

Figure 3. Yield of tomatoes in the Philippines, 1961–1965 to 1977. (Source: Bureau of Agricultural Economics, Philippines.)

Research and extension programs

In 1968 the National Food and Agricultural Council started a vegetable research and extension program at the University of the Philippines at Los Baños that led to the identification of varieties that were better adapted to local conditions. The program also provided training for research personnel in government and in private institutions. It continued until the creation, in 1972, of the Philippine Council for Agriculture and Resources Research, which developed a commodity research team on vegetables. Thus, there is now a body that conducts research on priority vegetables, of which tomatoes are one.

National vegetable production programs

A home garden movement, the "Share for Progress Project," was also created in 1968. It encouraged organizations and families to cooperate in using idle lots, labor, and other resources for productive purposes. The project later became the "Green Revolution Program," which included fruits and ornamentals as well as vegetables.

In 1975, as the counterpart of the national rice production campaign, a national commercial vegetable production, marketing, and credit program was launched. It is called Gulayan sa Kalusugan (GKP) or "Vegetable Production for Better Health Program." Production is encouraged by extending technical assistance to farmers, identifying suitable areas of production, helping to schedule planting and harvesting, and providing for efficient collection, distribution, and marketing. Under the financing arrangement that is part of the GKP, in each cropping season a farmer in the program can borrow up to US$1100 per hectare from a rural bank on easy terms.

Coordinated tomato production and marketing

An outstanding success of the GKP is the Coordinated Tomato Production and Marketing Program in Mindanao, about 750 kilometers south of the capital, Manila. From June to January tomato supplies are scarce in Manila, because of high temperatures, heavy rainfall, and frequent typhoons in most nearby tomato-producing areas. Farmers grow rice, the traditional rainy season crop, instead. But in northern Mindanao, at this time, climatic conditions are favorable for tomatoes. A few conscientious government officials and private citizens in that region had the foresight and initiative to organize a tomato production and marketing program.

Fresh-market tomato production began in northern Mindanao in 1968. As a public service, the Philippine Packing Corporation began teaching farmers in the province of Bukidnon to grow fresh-market tomatoes using the variety and methods the company had developed for producing processing tomatoes. The farmers were so successful that they decided to form the Mamdahilin Farmers Cooperative and to expand production.

Coordinating council. The Mamdahilin Farmers Cooperative, along with several independent operators, managed tomato shipping to Manila until 1975 when the Regional Coordinating

Council for the Tomato Industry (RCCTI) was established by the Bureau of Plant Industry to coordinate the production and marketing of fresh tomatoes. The RCCTI is chaired by the regional coordinator of GKP, with one member each drawn from the Trade Assistance Center, National Economic and Development Authority, Philippine Port Authority, Development Bank of the Philippines, traders, and producers.

Production strategies. The most important information needed in planning production was an estimate of the demand for tomatoes in metropolitan Manila, which has a population of nearly eight million. The estimate was obtained by reviewing the confidential files of tomato traders. After the trends were established, the figures were checked with the traders and the apparent demand was set at 375 tons per week from July to December.

To regulate the volume of production and stabilize the farm price of tomatoes, the concept of staggered planting was adopted. The goal of 375 tons of fresh-market tomatoes has been met by planting 18.75 hectares weekly from the first week of April to the first week of September. A total of 93.75 hectares is maintained per month to meet the target tonnage.

Even with careful planning and implementation, it has been necessary to employ some control mechanisms that are acceptable to both producers and traders. Producers must apply to the office of the Trade Assistance Center, RCCTI, and the office of the town mayor in the tomato production area for accreditation. Accreditation cards are issued to producers after a determination is made of their potential output and the general area where they grow tomatoes. The cards entitle holders to all benefits offered by the program. Then they are assigned volume allocations to be produced regularly each week.

Traders are required to submit tomato samples for inspection for chemical residues, insects and diseases, quality, and maturity. Only tomatoes that pass inspection receive shipping permits for entry and booking inside the Port Authority compound.

Collective marketing scheme. The RCCTI implemented the collective marketing scheme primarily to organize tomato producers for a joint marketing venture. Thus, if there are 10 groups each, with an allocation of 500 crates per week, the aggregate

In Claveria, Philippines, a farm family grades tomatoes before they are packed in wooden crates for shipment to Manila. (Source: RCCTI.)

total will be 5000 crates per week. Producers within a group are required to adopt a single farm plan and budget throughout the season. The operation is coordinated by a representative from the RCCTI (usually a technician from the Bureau of Plant Industry). Each group elects a trading partner who must (1) set a guaranteed farm price to producers plus suitable percentage rebates for whatever profits he may have for the group's transaction; (2) open a domestic letter of credit; and (3) liquidate accounts through a bank-to-bank transaction upon receipt of advice of collection from the producer's group.

The trading partners are likewise accredited with the Trade Assistance Center, where a marketing agreement is normally executed by both parties. Participating farmers benefit from the guaranteed price of farm produce, assured collection of accounts, patronage refund from bulk procurement of inputs, and close supervision of farm operations.

Financing operations. Three types of financing exist in the area: partnership, self-financing, and domestic letter of credit. Partnership financing is the most common. A trader, who is usually a wholesaler in Manila with business connections in northern Mindanao, advances money to the tomato producer for operational expenses in growing tomatoes. Acceptance of the money by the producer obligates him to deliver all his produce to the trader. Price fluctuations in Manila become the gauge of their pricing arrangements. In this scheme, the trader pays all expenses; the producer takes care of the tomato production. Profit sharing is based on gross sales at the wholesale price FOB Manila, with the trader getting 12 percent and the producer the remaining 88 percent.

A self-financing small trader (capacity of 2.5 to 6 tons per week) travels through the region buying tomatoes from groups of producers. Such a trader can offer a high farm price because he must acquire a sizable volume within a short time. Producers who have previous commitments sometimes sell to him because of the price he offers, but this type of outlet is unpredictable. The trader underwrites the cost of containers; of the labor for sorting, grading, packaging, and handling; and of transportation from farm to final market outlet in Manila.

Domestic letters of credit are used for financing by some

traders. A trader may open a domestic letter of credit with any commercial bank (principal bank) to finance his operation. The letter of credit can be liquidated within 90 days. With the letter of credit the trader can execute a marketing agreement with a producer in which a guaranteed farm price and regular weekly supplies of tomatoes are specified. When the trader receives tomatoes from the producer, he issues a delivery receipt and secures a bill of lading from the carrier. The trader gives the bill of lading to the producer, who may present it and the delivery receipt to an agent bank for collection of payment. The agent bank, in turn, collects the corresponding amount from the principal bank. In case of default, the trader loses his credit line and the bond he posted.

Progress and reasons for it

The program in Claveria started informally in 1973, although the RCCTI was not established until two years later. Within five years, yields in the supervised areas reached 20 t/ha (20,000 kg/ha), which is nearly three times the national average. The supervised area planted to tomatoes expanded from 188 hectares in 1973 to 375 hectares in 1977. During the same period the number of technicians assigned to help the growers doubled, from five to ten. The technicians believe that the yield plateau of 20 t/ha can be broken if better varieties and improved pest and disease control measures become available to growers.

Administrators, technicians, and growers attribute the progress of the program to government intervention, well-motivated field technicians, traders and growers who are willing to take the risks, favorable climatic conditions, and availability of appropriate varieties and production technologies.

Conscientious government officials stepped into the picture by coordinating the production and marketing of tomatoes. By controlling the issuance of permits to grow and market the produce, the government has been able to schedule production and reduce market gluts.

To monitor the different aspects of the industry, one field technician is assigned for every 40 hectares. Field technicians literally live at the farms they supervise. In general, they are properly motivated and adequately supported, receiving a token allowance of $14 per month, plus gasoline for their motorcycles, which have been obtained on an installment basis through the GKP.

Although some traders attempt to undermine RCCTI provisions by inducing growers to sell to them without contracts, most traders have been cooperative and have contributed to the progress of the program. In addition to handling the marketing of the produce, many traders advance production inputs to growers.

The participation of small farmers in the program has been crucial. Their willingness to take the risk of growing tomatoes during the rainy season makes the program possible. They have cooperated wholeheartedly with the technicians to follow production schedules and recommendations.

Claveria, located 920 meters above sea level, has favorable temperatures for tomato production (24 to 27° C. daytime, and 16 to 19° C. at night). Sunshine is not as abundant as it should be. Precipitation is normally high, but the rolling production areas permit good drainage.

Varieties and production technologies were transferred to Claveria from a nearby Philippine Packing Corporation farm. Without them, other efforts would have been futile. More recently a local government experiment station has been developing new varieties and management practices that will increase productivity in the region.

Another favorable factor is proximity to a port—Cagayan de Oro, 45 kilometers from Claveria—from which commercial ships sail to Manila two or three times a week and to an airline that has two scheduled flights daily between Cagayan de Oro and Manila. There are also good telecommunications. These services facilitate communications between the traders of the two cities and the transportation of perishable tomatoes.

TAIWAN

From 1951 to 1974, the area planted to tomatoes in Taiwan increased steadily—from about 1000 to 8000 hectares. Yields remained low until 1960, then began to rise, reaching 24.5 t/ha

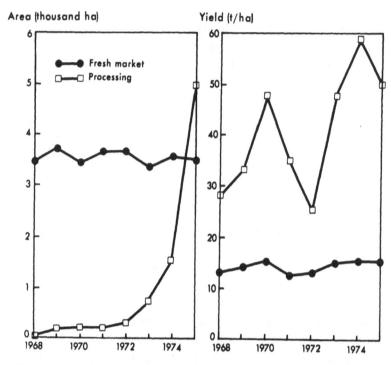

Figure 4. Area and yield of fresh-market and processing tomatoes in Taiwan, 1968–1975. (Source: Taiwan Agricultural Yearbook and AVRDC.)

in 1975, which represented a tripling of productivity within two and a half decades.

The rapid expansion of Taiwan's tomato industry was due mainly to the strong revival of the processing industry (Figure 4). While the area of fresh-market tomatoes was unchanged, the area of processing tomatoes rose from 18 hectares to 5000 hectares between 1968 and 1975. Fresh-market yields have been consistently lower than those for processing.

Climatic conditions

The climate of Taiwan is suitable for growing many crops. From October to March the warm days and cool nights are optimal for tomato production; consequently, Taiwan is the only country where processing tomatoes are grown on a large scale in the lowlands at this time of the year. Japanese and other temperate-bred varieties can be grown efficiently under such climatic conditions. Insect pests and diseases are less severe than they are during the rainy months. For example, water mold is not serious in the Tainan area, where about 80 percent of the processing tomatoes are grown, since tomato harvesting takes place between December to April, a dry period.

Irrigation facilities

Tomatoes require large amounts of water for growth and development, and it is adequately supplied when needed in Taiwan. Without irrigation it would be impossible for Taiwan to grow tomatoes at the time when rainfall is low or nonexistent.

Inorganic and organic fertilizer

Taiwanese farmers use both inorganic and organic fertilizers. On the average, per hectare, they apply about 490 kilograms of inorganic fertilizer and nearly 8 tons of organic fertilizer per year. For tomatoes they commonly apply 800 kg/ha of complete fertilizer (16-8-12), 100 kg/ha of nitrogen, 150 kg/ha of P_2O_5, 750 kg/ha of lime, and 15 t/ha of compost. Green manuring, composting, waste plant materials, and animal and human manures are widely used. It is common to see piles of organic fertilizer in the middle of a field as produce is being harvested. The farmer may dump a cartful of compost in the field and then leave the field with his cart loaded with produce. In the small villages nearly every house has its own compost pit and a pile of sugarcane and rice straw. Rice straw is seldom burned. It is also common to see bullock-drawn carts loaded with barrels of human manure.

Industrious farmers and small farm size

Taiwan's agriculture is more like gardening than farming. The intensive and continuous cropping, liberal use of fertilizers, irrigation, and efficient protection against crop pests are characteristic of a market garden. The industriousness of farmers is shown by their weeding rice fields on hands and knees, training sweet potato vines, watering individual vegetable crops with liquid manure, cutting individual maize stalks above the ear, and topping and pollinating individual tomato flowers.

The size of the average tomato farm in Taiwan was about 0.4 hectare in 1972. Because of their small farms, Taiwanese farmers are eager to improve productivity. They explore every possibility for obtaining extra income through intensive use of their limited land resources. The small farm size permits them to crop their areas intensively with family members as the main source of labor.

Infrastructure

Although the government of Taiwan does not give direct monetary aid to farmers, it has invested heavily to improve communication and transportation. It has also built small and large markets. Because of this assistance, farmers have been able to benefit from efficient production and marketing.

Growth of the processing industry

Taiwan's tomato industry began in the 1930s during the Japanese occupation. Capacity was limited and Japan was the main export market. After World War II, competition from the United States and Italy for the Japanese market resulted in the rapid decline of Taiwan's tomato processing industry. About 1967, however, some Japanese investors became interested in reviving the industry and established a cannery for tomato processing.

For five years crops were generally small. Farmers were initially hesitant about growing tomatoes under the unfamiliar contractual arrangements with the company. In 1973 bad weather in major tomato-producing countries such as Italy, Spain, and Portugal caused a sharp reduction in the world supply of processed tomato products. This opened the market for Taiwan. Within a short time, several factories were prepared to add tomatoes to their processing lines and they rapidly expanded their processing facilities. As a result, Taiwan's exports of canned tomato products soared from 4260 cases in 1967 to 1.6 million cases worth nearly US$19 million in 1976.

The chief reasons for the rapid expansion of Taiwan's tomato industry were strong export markets and an efficient system of procuring tomatoes from the growers (based on similar systems in Japan). The key ingredients were a well-trained and motivated technical assistance staff, the location of processing factories within potential tomato-production areas, and industrious farmers capable of shifting crops within their farming systems when given price incentives.

In general, tomatoes for processing are grown in three cropping periods. For early cropping, transplanting comes in the middle of September; transplanting for the principal cropping time occurs during the middle of October; and transplanting for late cropping is scheduled in November. Thus, harvesting of processing tomatoes starts in early December and extends until early May. Tomatoes planted earlier than September suffer from poor fruit setting, bacterial wilt, and leaf diseases such as leaf molds and gray leaf spot. On the other hand, tomatoes planted after December produce poor fruits due to sun scalding, blotchy ripening, and cracking. If these problems could be reduced, the production of tomatoes for processing in Taiwan could be extended.

Establishment of production goals

Plant processing capacity and expected export demand (advance orders) are used to determine the amount of tomato production needed. If bad weather is reported in major producing areas such as California (U.S.A.), production goals may be raised to meet additional demand.

A typical schedule prepared by a factory production manager for the production of processing tomatoes might be

May to June: Selection of farmer representative and signing
   of contracts
July to September: Sowing seeds in the nursery bed
October to December: Transplanting
January to April: Harvesting

Representatives of the tomato-processing companies usually meet each year during July to exchange information on the size of the contracted area and price. They also informally establish zones in which each company will concentrate its contract farmers to facilitate supervision and field visits by extension personnel.

Methods of collecting tomatoes for canneries

A major concern of canneries is to secure an inexpensive and reliable supply of processing tomatoes, especially in terms of quality, quantity, and time of delivery. They obtain their raw materials in various ways: unscheduled collection, farmers' association-factory contract, or a farmer-factory contract system.

Unscheduled collection. Small factories with limited production capacity and little investment in equipment and personnel buy tomatoes from local markets or even from farmers whose fields are contracted to other factories. Such actions disturb the distribution system and tend to result in poor quality canned products. To discourage them, the government now requires processors to establish a contract agreement in order to get an export license.

Farmers' association-factory contract. Sometimes a factory contracts with a farmers' association that in turn contracts with farmer-members of the association. The factory deals with the association directly to procure the tomatoes and payment is made through the association. Extension personnel from the association regularly visit growers' fields to ensure good production. For these services, the association gets $0.30 per 100 kilograms of tomatoes delivered to the factory. The factory provides crates for tomato collection and pays the cost of transporting the tomatoes from the field to the factory.

Farmer-factory contract. The most common scheme of tomato collection in Taiwan is the farmer-factory contract. Extension workers employed by the factory are sent to potential production areas. There they carefully select farmer representatives who organize local farmers, usually at the *Li* level (smallest political unit). The total land area committed by each group of farmers is approximately 20 hectares. Each farmer within a group contracts his crop to the factory in return for a guaranteed price, certain production inputs, and technical assistance.

Farmers enter into a contract with the factory. In most agricultural contracts in Taiwan, two guarantors must countersign for each production team member and be liable if the team member defaults. If the default occurs in supplying all tomatoes to the factory, the factory will cease purchasing from that farmer. However, if the default is for nonrepayment of production loans, action is taken against the farmer and guarantors. There are few defaults.

Terms of the contract include

- *Tomato variety.* The factory provides tomato seeds to the farmer representative, who plants them and distributes the seedlings to his production team for transplanting. This ensures that the entire team plants the same variety and that the quality and harvesting period at a given location is uniform.
- *Production inputs.* The factory provides certain materials, such as pesticides and fertilizers, to farmers and recovers the cost (plus interest) by deducting that amount from the value of the tomatoes delivered to the factory.
- *Contract price.* The farmers and factories agree upon a contract price prior to planting. In 1977/1978, for example, first grade was $3.60 per 100 kilograms and second grade was $2.60 per 100 kilograms. An incentive of $0.13 per 100 kilograms for planting an earlier variety or $0.08 for planting a later variety was paid to compensate for the greater risk. These additional plantings extend the factory processing period from December to May.
- *Transportation costs.* The factory usually pays transportation costs through the farmer representative at rates set according to the distance from the field nearest a highway or from the village collection area to the factory.
- *Product marketing.* The contract specifies that the farmer's total harvest must be delivered to the factory.
- *Grading.* The factory uses a random sampling method to determine the proportion of first-grade and second-grade tomatoes that farmers deliver each day. The grades are rigorously defined and mutually agreed upon, but grading

sometimes becomes a controversial point, especially when crates are delivered late or the factory delays acceptance of delivery. Payment is usually made two or three weeks after delivery.

• *Pesticide sprayers.* Sometimes sprayers are subsidized to help make organized pest control programs more effective. Most farmers own their sprayers, however.

Role of the factory. Through its extension staff, the factory organizes a scheme of tomato collection. Most of the extension workers are former staff members of the farmers' association or local government officials who know local farm conditions and can be trained in tomato production techniques. They select and supervise farmer representatives; coordinate the distribution of inputs to the farmer representatives; and provide technical assistance to tomato farmers on improved cultivation methods, insect and disease control measures, and the like (Figure 5). In addition to production inputs and technical assistance, the factory gives farmers a guaranteed price for their tomatoes.

Harvesting dates are regulated to keep daily supplies of tomatoes within factory capacity. However, during the peak harvest-

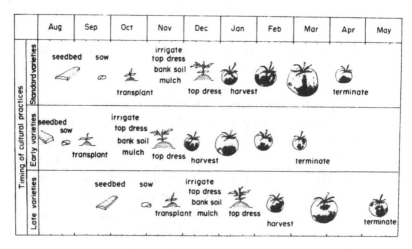

Figure 5. Timing of cultural practices for processing tomatoes in Taiwan: Example of a technical assistance aid supplied to contract farmers by the processing industry. (Translated from Chinese.)

ing period three work shifts are required to process the daily delivery of tomatoes. The factory, through the farmer representative, distributes crates to the production team members at harvest time and tomatoes are hauled to the factory by whatever means of transportation is available to each farmer. Near the factory ox carts deliver tomatoes from fields, but trucks of various sizes are used for longer distances. Because cold storage capacity is limited, open storage at the factory site is fairly common.

Other features. Farmers who participate in the collection scheme can either grow their tomatoes in a monoculture or intercrop them, usually with sugarcane. Monoculture usually gives higher yields than intercropping, but the total revenue from tomatoes and sugarcane is more profitable. Intercropping tomatoes with sugarcane is possible in Taiwan because the mild winter probably slows sugarcane growth. The sugarcane canopy does not overlap for four months so that tomatoes receive adequate sunlight. In other tropical countries such intercropping is not practiced since overlapping of cane canopy usually occurs in less than three months, making tomato harvesting difficult and lowering yields.

In the past, most varieties of tomatoes used in Taiwan were bred in Japan by the Kagome Company. Its subsidiary in Taiwan had exclusive use of new varieties during the first year of introduction, but the following year other factories were able to use them. Most factories do not conduct tomato research. They depend primarily on whatever Kagome Company can introduce. Recently, however, both government institutions and AVRDC have begun developing varieties for processing.

Another feature of the Taiwan processing industry is the availability of other crops—such as pineapple, asparagus, and mushrooms—that can be processed when the tomato season ends. Nearly every factory has high annual production capacity and year-round production assures factory workers of employment, which results in high morale and efficiency.

MEXICO

Between 1965 and 1975 annual mean tomato yields in Mexico doubled and the increase in the exports of fresh-market tomatoes

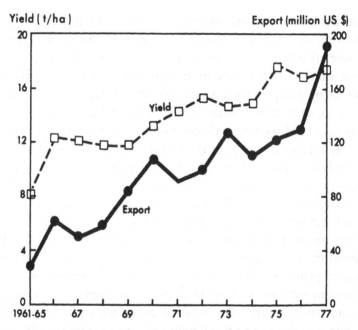

Figure 6. Yield and export value of Mexican tomatoes, 1961–1965 to 1977. (Source: FAO.)

was even more dramatic (Figure 6). In 1977, Mexico exported tomatoes worth US$193 million, an increase of approximately 450 percent in two decades. About four-fifths of Mexico's tomatoes go to the United States and Canada. The phenomenal increases in yield and in growth of the industry were brought about mostly by progress in fresh-market production. For instance, Mexican scientists report that yields of at least 25 t/ha are commonly obtained from large-scale production in the state of Sinaloa, where 40 percent of Mexico's tomatoes are grown.

Mexican scientists and tomato growers agree that the fundamental reasons for the successful production in Sinaloa are favorable climatic conditions, an excellent irrigation and drainage system, ready market outlets, and abundant, inexpensive labor. In addition, the modern cultural practices and postharvest operations that farmers use and the presence of organizations and cooperatives that provide services to growers make Sinaloa a progressive farming area.

Climatic conditions

Tomatoes thrive under a wide range of environmental conditions, but for optimum production they require plenty of sunshine, moderately cool night temperatures, warm days, and well-drained soil. They suffer from frost or prolonged chilling temperatures. The state of Sinaloa, particularly the Culiacan Valley, is an almost frost-free area. In Sonora (north of Sinaloa) frost is frequent during the winter months, and in Guadalajara (south of Sinaloa) there is too much rain at the time tomatoes for export to the United States should be grown.

In Mexico tomatoes should be planted from late August to February so that harvesting occurs from October to June, a time when outdoor tomato production in the United States is low. In Sinaloa at this time frost is rare, rainfall slight, and sunshine plentiful. As a result of the favorable climatic conditions growers can both achieve high yields and obtain high quality tomatoes.

Irrigation

It would be impossible to grow tomatoes without irrigation after October since rainfall steadily declines and becomes almost nil from February to May. During the dry season, the Culiacan Valley obtains water from dams and canals. In 1978, growers paid for the water at the rate of about US$35/ha/yr.

Market outlets

During the winter months, Mexico provides about a third of the fresh tomatoes consumed in the United States. (The rest come from outdoor production in Florida and from greenhouse production.) Some Mexican tomatoes also go to Canada during this period. The Culiacan Valley is the center of Mexican winter production, but from April to June, production in northern Sinaloa around the Guasave–Los Mochis areas supplements the declining volume from the Culiacan Valley.

Laborers irrigate a field of tomatoes in Mexico. Irrigation and water control are important for tomato production. (Source: CIAPAN.)

Inexpensive labor

Preharvest production costs are lower in Mexico than in either California or Florida, the major U.S. production areas, due mainly to cheaper labor. An abundance of inexpensive labor in Mexico permits most cultural operations and postharvest practices to be done manually instead of with mechanized equipment as in Florida. For a country with a high unemployment rate, this situation is favorable. However, as more factories are built in the rural areas and more jobs become available, it will be necessary to mechanize some farm operations.

Technical personnel

Mexico is developing a corps of trained researchers and extension personnel to help farmers in growing tomatoes. However, tomato operations in Sinaloa are so large (500 to 1000 hectares per operation) that growers can afford to hire highly qualified technicians to plan and supervise operations such as growing the seedlings, planting, water management, fertilization, and pest control. In addition, specialists are hired to manage the harvesting, packing, and transportation. Without them, it would be inconceivable to grow hundreds of hectares of tomatoes and market the produce in the United States.

New varieties

Mostly U.S. varieties are grown in Mexico, including Walter, Florida MH_1, Poleboy, and Tropic. Fusarium, a soil-borne disease, is a serious problem in Sinaloa, and Walter and Florida MH_1 have resistance to fusarium races 1 and 2. Poleboy and Tropic are resistant only to race 1. Walter constitutes at least 60 percent of the winter production.

The government experiment station in Culiacan—Centro de Investigaciones Agrícolas del Pacifico Norte (CIAPAN)—has released two hybrids (Bataoto and Buenavista), which the breeders report possess resistance to both races of Fusarium as well as having high yield potential. It remains for the government or private seed companies to make seeds of these hybrids available to growers.

Scientists at CIAPAN are developing varieties that can be grown as early as July or August so harvesting can be moved into late September or early October. Also, they plan to improve the resistance of varieties to virus and bacterial diseases such as bacterial canker and spot.

Growing seedlings

In 1972 a large grower in the Culiacan Valley demonstrated the advantages of using greenhouse-grown transplants in field production and now many other growers have begun to use this technique. At least 50 percent of the tomatoes and peppers shipped from Mexico are from greenhouse-grown transplants. Counter to the trend in the United States, it is expected that more Mexican growers will change from direct seeding to greenhouse-grown transplants.

The operation begins by germinating seeds in an artificial soil held in plastic foam trays. The seedlings are grown in large plastic greenhouses, where they are protected from the strong winds and rain that occur in August and September. In the greenhouse, the seedlings' growing conditions—soil moisture, humidity, fertilizer, and temperature—can be regulated and pests and diseases can be closely controlled. It is also easy to condition the plants and schedule the transplanting operations. Because transplants are being continuously produced, plants that are damaged in the field by winds or rain can be replaced immediately. The precise production and harvesting schedules needed to serve specific export markets can therefore be better maintained if greenhouse-grown transplants are used.

Transplanting

Four to six days before transplanting, a field that has been thoroughly prepared is irrigated. Mexican growers report that this procedure ensures the survival rate of transplants. All transplanting is done by hand. Under the crew approach commonly used in Culiacan to speed transplanting, more workers are assigned to making holes and planting than to distributing seedlings. On the average, a crew of 15 persons can plant 1 hectare a day. Mechanized transplanting is rare in Mexico because of the low labor costs.

Spacing

Seedlings are placed in rows with 15 to 25 centimeters between plants and 150 to 180 centimeters between rows. The widely spaced rows make field operations and harvesting convenient.

Fertilization

Most growers use soil analysis to determine the rate of fertilization for their fields. Many growers practice split application of fertilizer: Before planting, one third of the nitrogen plus all the phosphorus and the potassium, if any, are incorporated into the soil; 60 days after transplanting another third of the nitrogen is applied; 100 days after transplanting the final nitrogen application is made.

Staking

In the 1976/77 cropping season, 72 percent of the total area grown to tomatoes was staked. Although staking raises the cost of production, it also increases yields substantially, improves the quality of the crop, and makes harvesting easier. The increase in yield is attributed to better disease and insect control. Since application of insecticides, chelating agent, fungicides, and foliar nitrogen is usually done by airplane, leaves of the staked plants are more thoroughly covered by chemicals than those of plants that grow freely on the ground.

Packing and transporting

When the tomatoes are harvested they are taken to packing plants that have up-to-date equipment arranged in a large floor space. Tomatoes are dumped in the unloading section, washed, and then moved through a series of conveyors to be graded, culled, packed in boxes, air-blasted at 10° C. for about 3 hours, and then stored at 10° C. The following day the tomatoes are ready to be shipped in refrigerated trucks.

Boxes labeled with the name of the company, the grade, and the number of tomatoes in each box are assembled in one corner of the packing plant.

In spite of U.S. import duties, Mexican taxes, and high marketing and transportation costs, it is economical for Mexican growers to truck tomatoes a thousand kilometers to the U.S. border at Nogales, Arizona. The advantage of inexpensive labor, however, may be disappearing, and the imposition of a stiff tariff by the United States and prohibitive trucking costs could nullify the climatic advantage of Mexican production.

Cooperatives

Vegetable producers in the Culiacan Valley are well served by three cooperatives: Unión Nacional de Productores de Hortalizas, Confederación de Asociaciones Agrícolas del Estado de Sinaloa, and Asociación de Agricultores del Río Culiacán. These cooperatives are self-financed through deductible payments on a kilogram or a package basis. They assist in all agricultural operations through involvement in the purchase of farm equipment, seeds, fertilizers, and chemicals, as well as in the professional training of researchers. They arrange insurance and services for marketing tomatoes. Without these cooperatives it would be difficult to have successful tomato production in that part of Mexico because the government has not been active in encouraging vegetable production for export. So far the industry has grown mainly as a result of private endeavor.

Government

Although the Mexican government has not been involved in the industry directly, it has invested heavily in means of communication, roads, and irrigation projects. It has also helped by improving the research and extension capabilities of Mexican scientists. A multidisciplinary team of scientists is conducting tomato research at the government-supported station CIAPAN.

3
Guidelines for
Program Development

Although research on tomatoes should be undertaken at the national level, a production program for tomatoes need not be nationwide. Commercial production should take place in the areas where most ingredients for success exist. In fact, successful tomato production around the world has generally been concentrated in areas where climate is particularly favorable.

In Mexico, for example, Culiacan, Sinaloa, has the best climatic conditions for winter tomato production and the best irrigation facilities. Similarly, California produces about 80 percent of the total processing tomatoes grown in the United States because long, dry summers and a well-developed irrigation system throughout the central valley allow farmers to grow tomatoes much more efficiently than farmers in other parts of the country. Florida, on the other hand, is the only area in the United States that produces tomatoes outdoors in the winter, and it grows 45 percent of the nation's outdoor spring production because of favorable climatic conditions.

Regardless of whether tomato production is regional or national, a concerted, unified effort among government officials, scientists, extension workers, private businesses, and farmers is essential. A coordinating body can be established to bring these groups together. One of the first actions of such a body should be to set up a study committee to analyze the domestic and export potential for tomatoes. Alternatively, an existing government or private organization or a national food production program may be able to do the job.

Analyzing the demand

Domestic demand. There are at least two ways to determine local demand for fresh tomatoes: study the monthly volume of tomato transactions in major markets or establish consumption patterns and per capita consumption. A combination of the two methods will probably yield a better estimate.

The volume of monthly transactions provides an indication of the demand and the yearly fluctuation of supplies. Such information can be used to estimate how much to produce each month. This method was used by the government-supported tomato program in the Philippines to establish the apparent demand in Manila. Initially, the monthly volumes were obtained from traders who kept records of their transactions. Later, estimates were substantiated by the bills of lading issued by the carrier that transported the tomatoes to Manila. A similar technique will probably work in any area where there are reliable traders who keep records of their sales. In Taiwan, the volume of transactions in any vegetable is systematically recorded at every wholesale market.

If consumption data are available, they can be useful for estimating the domestic demand potential. Knowledge of per capita consumption can be used to calculate a rough estimate of potential demand. Familiarity with consumption patterns regionally and by income group can indicate what sort of demand potentials to expect in rural and urban households. For example, studies in Brazil established that per capita consumption of tomatoes varied from about 3 kilograms in the northeastern region to 9 kilograms in Sao Paulo.

The requirement for processed or canned tomatoes can be established by checking import statistics maintained by the government import/export bureau. The quantity imported plus local production will constitute the apparent demand for processing tomatoes.

Export demand. Developing countries favored by exceptional climate and proximity to market can export reasonable quantities of fresh-market tomatoes to developed countries, thereby increasing farmers' incomes and adding to the country's foreign exchange. Information on the potential export demand can be obtained from business contacts or from records of the import/ export bureau. In Taiwan, for instance, traders from Hong Kong and Singapore arrange with the Provincial Farmers' Association for delivery of a specific volume of fresh-market tomatoes over a specific period of time. The provincial association, in turn, allocates the contracted production volume among the various local farmers' associations. The export demand to be satisfied in this case depends on the volume traders can handle. In Mexico, demand for export of fresh-market tomatoes is determined largely from export records and market information gathered by the cooperatives that service growers.

Commercial attachés of both exporting and importing countries and business contacts in the importing countries are good sources of information on potential export demand. For large and established factories, however, a market survey must be undertaken and buyers in the importing countries must be sought. Food factories in Taiwan regularly send their marketing staffs around the world to find buyers for their products. Advertising and displays of products in international expositions help to interest prospective buyers; supermarket chains in developed countries are good outlets for processed tomatoes from the tropics.

Analyzing available resources

In addition to establishing the demand potential, it is advisable to thoroughly analyze the natural, human, and financial resources of the region or country. Such an analysis can be effectively done by a team consisting of at least three experienced specialists: a horticulturist, an economist, and an irrigation specialist. The addition of individuals from other disciplines would strengthen the team.

Natural resources. Economists generally use the principle of comparative advantage in deciding what should be produced in a locality. The principle is based on the premise that each area should produce those products for which it has the greatest ratio of advantage as compared with other regions. This is where the availability of water and suitability of climate and soil conditions come into the picture.

Usually for a secondary crop like tomatoes, the availability of irrigation facilities hinges on the extent of government-supported irrigation projects. Information on a country's water resources can probably be obtained from government agencies, or a survey should be arranged through local or international agencies or a commercial consulting firm.

Because tomatoes cannot tolerate waterlogged soil, production areas should be well drained or provision should be made to construct suitable means of drainage. For rainy season production, rolling land where drainage is not likely to be a problem is better for growing tomatoes than flat lands.

Tomatoes thrive when there is plenty of sunshine, moderately cool temperature at night (15° to 20° C.), and warm daytime temperatures (25° to 30° C.). The optimum night temperatures should occur at the time the plant is flowering. Frequent strong winds and prolonged rainy periods and cloudiness are not good for tomatoes. Strong winds damage branches and flowers, and prolonged rainy periods predispose tomatoes to attacks of various diseases. Extensive periods of cloudiness retard photosynthesis and increase the susceptibility of certain varieties to root and leaf diseases. Therefore, long-term weather data (minimum, maximum, and mean temperatures; rainfall; solar radiation; daylength; and wind speed) are useful for planning and implementing the program.

If water resources and climate are satisfactory, soil conditions are usually not a limiting factor in choosing a site for tomato production. In many instances, anchorage is the primary function of soil in vegetable production. For example, in the Philippines' Mountain Province, growers use tons of chicken and cow manure and commercial fertilizers to improve the poor soil of their farms. Taiwanese farmers also use those soil amendments, plus a green manure crop every year or so. In the United States, marginal and rocky land, with appropriate fertilizers, has long been used for fresh-market tomato production in south Florida. Such soil improvements may not be sufficient for some problem soils because they would be too expensive and take too long.

Human resources. The programs discussed in Chapter 2 each involved a group of dedicated agricultural leaders, adequately supported and motivated extension personnel, industrious small farmers in Taiwan and the Philippines, and experienced hired labor in Mexico. The availability of good managers and skilled laborers in the locality should be assessed.

While leaders who will plan and implement the program ought to know local conditions, leaders can, if necessary, be brought in from other areas. Similarly, production and factory managers can be imported. In Taiwan, for example, most farm and factory managers were formerly Japanese specialists. Now, however, most of the key positions are held by Taiwanese. Some of the large tomato-processing plants in Africa and Latin America are being manned by expatriate specialists, but as more nationals become trained they will eventually take over those jobs. Thus, key personnel to run the operations can be imported from developed countries at first and local leaders trained to do the job later.

Adequately trained extension specialists are urgently needed in developing countries. The number required for an individual production scheme depends on the area to be covered, how scattered the producers' farms are, and the transportation extension personnel will use. In Claveria, Philippines, a tomato extension specialist who owns a motorcycle can effectively cover 40 hectares of production. If the specialist had to rely completely on public transportation, he could handle only about 5 hectares.

The number of farmers who are willing to participate in the program and the number of laborers who can be hired (permanently or temporarily) also must be determined. Some farmers who are in the most suitable tomato production area may not be interested in growing tomatoes following the staple cereal crop. Lack of laborers can be a serious limitation too, especially in areas far from population centers. Sometimes enough laborers are available but they are of poor quality—they work only four out of eight hours and are unwilling to learn field operations. These situations should be considered in assessing human resources. The prevailing pay rates for hired labor and the availability of other jobs in the same general area that will compete with the operation also should be determined.

Financial resources. There are three probable sources of financial assistance for a tomato-production program: banks, cooperatives and farmers' associations, and traders. Preferably, all three should be available before a program is organized in an area.

Growers in a government-supported production program usually can obtain loans from government agencies more easily than from private banks. Government banks tend to be more lenient in terms of collateral and repayment schedules and they generally charge lower interest rates. The extent of cooperation from both types of banks to facilitate loan releases to farmers must be determined by whoever will implement the program. Possible arrangements to release to farmers a portion of each approved loan in kind (seeds, fertilizers, pesticides, etc.) should be explored for tomato production.

In Taiwan farmers usually obtain production inputs from cooperatives and farmers' associations in which they are members, rather than from government or private banks. Also, factories provide inputs and deduct the cost from payments after the tomatoes are delivered. In both developed and developing countries, cooperatives and farmers' associations provide other services in addition to financial assistance. If the prospective production area lacks such organizations, their establishment should be vigorously pursued.

The close relationship between traders and growers in Claveria, Philippines, permits financing of the tomato-production operation. There, a mutually beneficial association is encouraged by the government since both parties are satisfied with the arrangement. Cash advances are provided by the traders to the growers with the stipulation that all the produce will be delivered to the trader at harvest time. Price fluctuations in the market guide the pricing arrangements. Cash advances are liquidated after profit sharing based on gross sales at wholesale prices. The producer gets 88 percent and the remaining 12 percent goes to the trader, in contrast with some situations in which traders who advance inputs end up with the major share of the profits. If other financing methods are not available, this method should be explored.

Implementing the program

Once a decision has been made to go into production (based on the results of analyzing the demand and available resources), implementation of the program is left to the production manager or a coordinating council. Implementation should be coordinated so that the purchase of inputs, production of tomatoes, and flow of tomatoes to the markets or factories is done systematically. A good plan sometimes fails because of poor coordination and lack of effective leadership.

Setting production goals. The analysis of local and export demand should provide the basis for production goals. A schedule of deliveries also can be helpful in planning production. Adherence to production goals is especially important for fresh-market tomatoes because of their perishability. Production schedules for processing tomatoes should be well planned, too, so that deliveries to the factories can be processed at once and need not be stored in the open for long periods.

For small growers who will participate in the production program, it may not be difficult to follow schedules precisely, but if the operation involves several hectares the first year should be considered a trial run for everyone. If possible, the program should start with small areas and expand later as the crew gains experience. Problems in the growing of seedlings, control of pests and diseases, irrigation, fertilization, and harvesting become compounded as the area grows larger. It is common to have a glut of ripe tomatoes in the field, but losses due to overripe tomatoes can be minimized if planting is staggered to fit established production objectives.

Training extension staff. Adequate training and experience are the basis for developing effective personnel who will transfer technology to growers. Local training in a tomato-production program elsewhere in the country would be ideal. Inviting scientists from abroad to conduct training is also a good way to get started—many people can be trained simultaneously this way and it may be cheaper in the long run. On the other hand, if the foreign trainer has not had tropical experience he may spend much of his time learning the local problems. If the country does not have the capability to train people for tomato production, carefully selected individuals should be sent abroad. Whenever possible trainees should be sent to a developing country that has advanced tomato production and in which the environment and culture are like those of the home country. Training in a developing, rather than a developed, country serves two important objectives: the training is more relevant, which results in a quicker transfer of technology, and there is less temptation to stay abroad.

Practical experience in growing the crop is complementary to appropriate training. At AVRDC a tomato production trainee spends four to five months performing all the cultural operations, from sowing to harvest. Such practical experience gives him confidence when he makes recommendations to tomato growers. As soon as trained personnel return to their posts, they can start training other individuals who will serve as extension specialists. (Appendix A lists places where research and production training can be obtained.)

Organizing farmers. Farmers' organizations in many forms—communes in China, *kibbutzim* in Israel, *Saemaul Undong* in South Korea, cooperatives in Mexico, farmers' associations in Taiwan—have been responsible for raising crop productivity in numerous countries. The Collective Marketing Scheme in the Philippines has resulted in excellent progress, as have the farmers' associations and contract agreements in Taiwan.

Benefits from such organizations vary. For example, those in Taiwan conduct marketing and extension activities, extend credit, and supply inputs, consumer goods (rice, sugar, canned foods, toiletries, etc.), and even luxury items such as color television sets and refrigerators. Some processing companies and wholesale buyers also assist in supplying inputs at reduced prices.

Experiences in both developed and developing countries clearly show the importance of involving the farmers themselves in crop production and rural development programs. This is best accomplished through nonpolitical farmers' organizations. Organizing farmers is critical for perishable commodities such as fruits and vegetables. A subsistence grower can handle his production without much of a problem, but when he starts to market some of his produce he needs various services that can best be handled by an effective organization. Such an organization should be run by competent, well-trained, well-paid, and honest people. All too often failure of organizations is due to embezzlement and poor management. Most developing countries need farmers' organizations on a village, regional, and national level.

Training people to run these organizations should be given high priority.

Availability of inputs. Many farmers in developing countries do not use fertilizers or pesticides because these inputs are too costly or are not available when needed. In successful tomato programs, managers see to it that production inputs are available at the right place and at the right time and, when necessary, arrange for timely release of bank or cooperative loans or extend part of the loans in kind (as fertilizer, seeds, chemicals, etc.).

A farmer-factory contract in which the farmer gets a fair price for his produce is a favorable mechanism for both parties. The factory advances the inputs and deducts the equivalent amount after the grower has delivered the produce. If the crop fails, repayment can be deferred until the next crop. In the Philippines, funds for fresh-market tomato production are commonly obtained through a trader who is willing to invest in a profit-sharing venture. Cash advances are made to the growers and repayment is made at the end of the season. In addition, landlords and private moneylenders also can provide loans to growers.

In each of these arrangements, repayment is more or less guaranteed. This is generally true for farmers' associations, too, since only members can avail themselves of credit. As a member of the association, it is the farmer's duty to repay his debts promptly to maintain the viability of his association. Whether or not credit needs of small tomato growers in the tropics can be met in similar ways, government-lending institutions should, in addition, develop viable credit systems for them. Growers, on the other hand, should be informed about the value of repayment of loans and promptness in doing so. The rate of loan repayment has been disappointing in many projects in developing countries.

Price incentives. In some countries governments support the prices of agricultural commodities, subsidize inputs, or both. Although no such policies have yet been adopted for tomato production, governments may want to consider including tomatoes in price incentives to encourage production. Factories and farmers' associations can sell inputs at lower prices because they buy them wholesale. Furthermore, canning factories usually offer a guaranteed price for tomatoes. This gives the grower assurance of a market for his tomatoes no matter how large his production, although the price may be slightly lower than the prevailing price in the markets.

It is more difficult to guarantee the price of fresh-market tomatoes than that of processing tomatoes, because the former are more subject to supply and demand fluctuations. Nevertheless, some enterprising traders from Hong Kong and Singapore have been giving a guaranteed price for Taiwan's fresh tomatoes during the summer season. Moreover, to encourage increased production, the traders review and adjust the guaranteed price in consultation with the officers of the farmers' associations. Growers who are guaranteed a fair price for their crop will deliver the desired quality and volume to the best of their ability.

Such a pricing policy for tomatoes may be applicable in other developing countries. Any regional or national tomato production program should find ways to introduce reasonable price policies that will benefit growers, traders, and consumers.

Maintenance of the program

Food production programs in developing countries often slow down after initial success. Why? Among the reasons are complacency among participants, resignation of staff members, shifting of emphasis to other programs, and loss of farmer interest. Agricultural leaders, factory managers, traders, growers, and extension personnel agree that initial success in tomato production can be maintained and even expanded through continuing research and extension support, frequent contact between extension personnel and growers, and vigorous campaigns to expand the market.

Staff morale. If the morale of personnel who conduct research and extend information to farmers is poor, inefficiencies and resignations result. Unequal treatment of research and extension staff is a common cause of low morale. Insofar as possible, these groups should be accorded equal pay and opportunities. Other means of maintaining staff morale are adequate funding, regular inservice training, and proper incentives.

Adequate funding encourages staff members to stay on the job and to be achievers. When there are insufficient funds for experiments and trials or for developing extension materials staff members must spend time looking for money instead of carrying out their principal functions. Some staff members prefer to stay in their offices rather than solicit financial support. Neither situation is productive.

Staff members should also be able to take part in inservice training, meetings, and other opportunities for professional advancement. Trips to meetings outside the country should be allowed at times to add prestige, boost morale, broaden perspectives and contacts, update knowledge, and sharpen imagination. Many ideas and practical innovations can be gained from these activities. More frequent professional contacts and exchange of information are badly needed among the staff members of government and private agencies in developing nations. In the absence of formal meetings, frequent interactions among scientists of different disciplines and extension staff members through visits and consultations should be encouraged and adequately funded.

Various incentives for exceptional individuals have been used in food production programs in developing countries. Some of the incentives are cash, public recognition, increase in salary, advancement in rank, scholarships, trips abroad, and an audience with the president or prime minister. In any form, incentives give recipients confidence, make them realize the importance of their roles in the program, boost their morale, and encourage them to be even more effective. Moreover other staff members see that they have the opportunity to be similarly rewarded.

Importance of research. Despite adequate training and proper motivation, extension personnel may be ineffective if they have few answers to growers' questions. They need the assistance of scientists who are investigating production problems and developing new varieties and agricultural technology.

Research on new varieties and technology is best carried out in locations where the results are likely to be used. A good variety in one location may be a failure in another; therefore, it is important to test any variety or new technology locally before making firm recommendations. Nevertheless, findings from other locations can serve as models for solving certain problems. In exceptional cases where the variety or new technology is the only one available and is markedly superior to the existing one, an interim recommendation can be made.

Research is a continuing process. One problem may be solved, but a new one always appears. For example, the breeding of disease-resistant tomatoes has increased yields, but the contest between pathogens and scientists continues as new strains and races of diseases arise to replace those brought under control. And there is always room to improve current management practices to increase yields still further. For these reasons, successful food production programs give vigorous support to research. In the United States, local growers and canners provide regular financial support to the scientists at the University of California to carry out tomato breeding, evaluation trials, and research on grading, grade standards, and inspection procedures. In developing countries, if the government does not provide support to the industry, the private interests through the growers should pool their resources and support scientists to conduct research for them.

Contacts with farmers. Teaching farmers how to use new technologies is best done by extension specialists through on-farm trials and frequent field visits. Another method of transferring technology to farmers is through the mass media.

Extension specialists carry out trials of new technologies (varieties and practices) on farmers' fields and conduct field days to which local farmers are invited. Although such trials can be conducted in government experiment stations, it is advantageous to place them in farmers' fields. When farmers are directly involved they become more quickly convinced of the productivity and profitability of new ideas.

In the Philippines and Taiwan, extension specialists visit farmers frequently to advise them on promising varieties and appropriate technologies and to help schedule planting and harvesting operations. They are on constant call so farmers can rely on them. Extension men in Claveria, Philippines, practically live on the farm. To do all these jobs, it is essential for extension workers to have a dependable means of transportation. In

Claveria, extension specialists are provided motorcycles.

Ideas tested in campaigns to raise cereal production in various countries might also be adopted by extension specialists who work with tomato growers. In some areas, extension workers help farmers make loan applications and help collect the loans when they mature. To promote awareness of new techniques, they write for newspapers, appear on radio and television, and distribute leaflets and booklets to farmers. Establishing prizes and awards for outstanding farmers can help publicize successful management practices.

Expanding market outlets. As production increases there is a need to expand market outlets locally or abroad. The most common methods used involve intensive local advertising and the development of new products and ways of using tomatoes.

Another method is to send trade missions abroad. Private companies in Thailand and Taiwan send trade missions to Europe, the Mideast, and North America to find markets for their agricultural products. The Taiwan government encourages factories to display their products in consulates and embassies to acquaint potential buyers with these products. Regional trade expositions are periodically sponsored by the Association of Southeast Asian Nations at which all kinds of products, from agricultural produce to equipment, are displayed. Similar efforts by other developing countries can expand market outlets.

Unless soils and climates are *especially* suitable for tomatoes, however, production should be expanded primarily to improve national diets rather than in the expectation of earning foreign exchange through exports. In the intense competition for export markets, the advantages of naturally favored tomato production areas are overwhelming.

Local consumption will increase if tomatoes are regularly available in the market at an attractive price to consumers. Even with minimum advertising people buy tomatoes if they are available at reasonable prices. More efficient production will also result in cheaper processed products if costs of tin cans and labor do not increase drastically. Development of new products acceptable to consumers and of new ways of using tomatoes will surely improve demand.

Part 2

Scientific Considerations

4
A World Crop

Tomatoes have become one of the most popular and widely grown vegetables in the world. Until the nineteenth century, however, the tomato was grown chiefly as an ornamental plant, for its colorful fruit. As a food, it was avoided in the belief that its relationship to poisonous members of the nightshade family (such as belladonna and mandrake, which contain high concentrations of alkaloids) made it toxic. It is now known that the predominant alkaloid in tomatoes is tomatine, a much less toxic compound, even at high concentrations, than the alkaloids of most other nightshades. Nevertheless, some people in countries such as Sri Lanka, India, and the Philippines believe that eating too many tomatoes can cause severe stomachache. In the rural Philippines, many villagers claim that swallowing tomato seeds can cause appendicitis, but on the other hand many think that mothers-to-be who eat a lot of red, ripe tomatoes will have babies with fair complexions and rosy cheeks.

Origin and distribution

Although the tomato's origin and the early history of its domestication are obscure, the weight of evidence suggests that Mexico was the probable center of origin. Three aspects of its origin are reasonably certain: the cultivated tomato originated in the New World; it reached a fairly advanced stage of domestication there before being transported to Europe and Asia; and its most likely ancestor is the wild cherry tomato (*L. esculentum* var. *cerasiforme*) found first throughout tropical and subtropical

America and then in the tropics of Asia and Africa.

All tomato varieties now grown in Europe and Asia are descendents of seeds carried from Mexico to Europe and Asia in the sixteenth century by European merchants and colonizers. African tomatoes, however, probably descend from varieties introduced later from Europe.

Historical records show that tomatoes were taken to Europe by Cortez in 1523, soon after the conquest of Mexico City. However, the earliest mention of the existence of tomatoes in the Old World was made in 1554 by an Italian herbalist, Pier Andrea Mattioli. As for Asia, the Spanish began introducing several agricultural commodities into the Philippines from Mexico in 1571, but it is possible that tomatoes had been taken from Spain to Asia much earlier, perhaps just a few years after the discovery of the Philippines by Ferdinand Magellan in 1521. Trade between the Philippines and the neighboring countries of China, Japan, and India may have been responsible for the spread of tomatoes into those countries. It is also possible that the British, Dutch, and French encouraged the introduction of tomatoes into their Asian colonies.

Production areas

Tomatoes thrive at many latitudes and under a wide range of soil types, temperatures, and methods of cultivation. They have been productively grown outdoors from the equator to Rio Gallegos, Argentina (at 52 degrees south latitude), and to Edmonton, Canada (at 54 degrees north latitude). Plant explorers have found wild relatives of the tomato in the tropical rain forests of South America as well as in arid regions of its native Mexico.

Provided with adequate nutrients, cultivated tomatoes can be produced commercially in water culture (hydroponics) and in gravel or rock cultures. They are, however, sensitive to lack of oxygen in the root zone caused by poor drainage. Therefore, tomatoes grown hydroponically must be supplied with oxygen in the water in addition to other nutrients.

There are breeding lines of tomatoes that grow and produce fruits equally well under temperature extremes, but the plants suffer from frost or prolonged exposure to temperatures below 10° C. For this reason, even in the tropics, large-scale outdoor tomato production is rare at altitudes above 2000 meters, where frost and chilling temperatures are likely to occur.

In the tropics, tomatoes are grown in both the highlands—areas 500 to 2000 meters above sea level—and the lowlands—areas less than 500 meters above sea level. Because of the influence of precipitation, the highlands can be either cool-dry or cool-wet. The lowlands, on the other hand, may be cool-dry, hot-dry, or hot-wet.

Highlands. Tomato production in the tropics tends to be more successful in highland areas, primarily because of mild temperatures. In most tropical Asian countries, the principal drawback of highland tomato production areas is their distance from the major cities or population centers. With poor transportation and inadequate packing, losses in quality and quantity

Tomato production near Palmira, Colombia: In the tropics, tomato growing is usually most successful in highland areas. (Source: Instituto Colombiano Agropecuario.)

before tomatoes reach market are substantial. Losses are less serious in Latin America because the highlands are usually closer to the cities than they are in Asia.

Provided that no frosts or chilling temperatures occur, the cool-dry season of the highlands is the best for tomatoes. Temperatures are optimal and sunshine is adequate. The dry fields facilitate land preparation and other operations such as cultivation, harvesting, and control of insects, diseases, and weeds. Moreover, insect pests, weed problems, and diseases are less hazardous at this time. However, if rainfall is inadequate for tomatoes, a source of irrigation water is needed.

During the cool-wet season in the highlands too much rain, poor drainage, and insufficient sunshine can be serious problems for tomatoes. The disadvantages of excessive rainfall and overcast skies with low light intensity will usually outweigh the advantage of favorable temperatures for the growth and development of the plant and for fruit setting. Under these conditions, moreover, many leaf diseases attack tomatoes, lowering yields and quality. Nevertheless, many farmers plant tomatoes during the cool-wet season because of the good price they bring on the market. For instance, in Claveria, the Philippines, about 500 hectares are grown outdoors during this season every year for shipment to Manila, 750 kilometers away. In Bogota, Colombia, and in Baguio, the Philippines, several hectares of tomatoes are produced during the cool-wet season, but under plastic houses as protection against the high rainfall. Although this system is costly, growers have found it profitable. In Latin America—particularly in Colombia, Brazil, Costa Rica, and Ecuador—fresh-market tomatoes are generally grown in the highlands.

Lowlands. In tropical Asia, population densities are greater in the lowlands than in the highlands, which suggests that vegetable production should be developed in the lowlands. Most vegetable production is in the highlands, although there is less land there. Production in areas closer to population centers would assure consumers of better quality vegetables than those that come from distant farms. It could also mean enormous savings in fuel and energy. The new industry would create employment since vegetable growing, particularly tomato production, is labor intensive. When it became necessary to mechanize field operations, the flat lowlands could be worked much more easily than could rolling or terraced highlands. In the future, the tropical mountains or highlands could be used for permanent crops such as fruit trees that require cool temperatures.

In the lowland tropics, tomatoes are mostly grown during the winter when climatic conditions are favorable (near optimum temperatures and low rainfall). Successful production is assured in areas that have good irrigation and drainage systems. In much of the Asian tropics, this season starts between October and November, when cool air from Siberia comes down into the region and remains until March. Thus, planting coincides with the time when rainfall is low and plant growth and development and fruit setting take place when temperatures are mild. The peak harvesting season in this region is from February to April.

In Mexico the lowlands of the Culiacan Valley have a similar cool season brought about by winds from the Sonora desert. In this season, as in the cool-dry season of the highland areas,

Tomato production in Senegal: Growing tomatoes in tropical lowlands is difficult but advantageous because population centers usually are nearby. (Source: Center for Horticultural Development, Senegal.)

insect pests, diseases, and weed problems are less severe than they are in warmer periods. High yields and low-cost production make it the best time to produce tomatoes for processing.

Conditions in Bangkok, Thailand, from February to April typify a hot-dry season. Temperatures rise to 40° C. or more during the day and usually fall no lower than 25° C. at night. At these temperatures, most tomato plants fail to set fruit, apparently because the pollen bursts. Evaporation and evapotranspiration are so rapid that plants become dehydrated and wilt, especially if no supplemental irrigation is provided. Because the whitefly population increases at high temperatures, leaf curl virus, which is transmitted by the white fly, often severely damages tomatoes grown at the beginning of the hot-dry season. Sun scalding of fruit is also common at this time.

The hot-wet season is the most difficult time to grow tomatoes in the lowland tropics because of excessive soil and air moisture, unfavorable temperatures and conditions for fruit setting, and rapid development of pests, diseases, and weeds. At this time, the supply of tomatoes in the market is low and consumer prices are high. Despite the difficulties and high production costs, some enterprising growers in the tropics choose to produce fresh-market tomatoes during this season in order to take advantage of the prices.

Increased production

The impact of the "green revolution" has prompted policy-makers and administrators in many developing countries to take an interest not only in cereals but also in other crops, such as tomatoes.

Data from two contrasting periods are presented in Table 5 for the area, yield, and production of tomatoes. The 1961–1965 period represents a time when developing countries paid little attention to food crops other than cereals; the 1973–1977 period was a time when intensive food production programs in several developing countries began to include such vegetables as legumes and tomatoes. Between the two periods, tomato yields rose by about 3 t/ha in tropical countries and by 5 t/ha in temperate countries.

Table 5
Average annual area, yield, and production of tomatoes in tropical and temperate countries, 1961–1965 and 1973–1977

	1961–1965			1973–1977			Increase or Decrease (%)		
	Area (thou. ha)	Yield (t/ha)	Production (thou. tons)	Area (thou. ha)	Yield (t/ha)	Production (thou. tons)	Area	Yield	Production
TROPICAL									
Africa									
Cameroon	6	1.7	10	8	1.8	14	33	5	40
Ghana	3	5.7	17	21	4.7	98	600	– 18	475
Nigeria	19	9.8	186	25	9.0	225	32	– 8	21
Sudan	8	12.4	99	12	11.6	138	50	– 6	39
Tunisia	9	10.7	96	16	15.6	249	78	46	159
Asia									
Bangladesh	8	6.2	50	8	7.0	56	0	12	12
India	53	9.3	495	72	9.5	681	36	1	38
Indonesia	41	5.8	240	59	6.3	372	44	8	55
Malay Peninsula	3	5.0	15	5	5.4	27	67	8	80
Philippines	16	3.4	54	19	7.0	132	19	106	144
Saudi Arabia	6	12.8	77	17	13.4	227	183	4	195
Sri Lanka	6	1.8	11	5	3.0	15	– 17	64	36
Taiwan	3	9.0	27	7	21.1	146	133	134	441
Thailand	3	3.7	11	5	3.0	15	67	– 19	36

Table 5 (continued)

	1961–1965			1973–1977			Increase or Decrease (%)		
	Area (thou. ha)	Yield (t/ha)	Production (thou. tons)	Area (thou. ha)	Yield (t/ha)	Production (thou. tons)	Area	Yield	Production
Western Hemisphere									
Bolivia	4	13.8	55	5	10.2	51	25	-26	- 7
Brazil	36	13.9	502	47	22.9	1070	31	64	113
Colombia	3	13.7	41	7	20.3	141	133	49	244
Cuba	19	5.0	96	24	7.2	174	26	42	81
Guatemala	6	6.7	40	10	7.3	74	67	9	85
Mexico	58	8.0	465	61	16.6	1011	5	107	117
Peru	3	11.3	34	6	11.7	70	100	3	106
Venezuela	4	15.8	63	5	19.2	96	25	22	52
TEMPERATE									
Africa									
Algeria	8	15.5	124	12	10.6	124	50	-31	0
Egypt	72	14.8	1069	130	15.2	1976	81	2	85
Libya	5	11.0	55	16	11.6	186	220	6	238
Morocco	13	17.9	233	14	28.3	396	8	58	70
Asia									
China	211	10.5	2217	201	16.1	3241	- 5	54	46
Iran	16	7.9	127	22	10.6	233	38	33	84

Iraq	24	6.7	161	37	8.9	331	54	33	106
Jordan	21	9.7	203	11	9.9	107	− 48	3	− 47
Lebanon	3	12.3	37	5	13.4	67	67	9	81
Syria	17	7.8	133	29	14.5	421	71	85	216
Turkey	57	21.0	1199	82	29.3	2410	44	39	101
Western Hemisphere									
Argentina	19	16.3	310	30	18.2	545	58	11	76
Chile	8	19.5	156	7	24.3	170	− 12	25	9
U.S.	178	28.5	5079	187	39.4	7402	5	38	46
Europe									
Greece	28	14.6	408	37	37.5	1387	32	157	240
Italy	126	22.8	2875	110	30.4	3346	− 13	33	16
Netherlands	3	85.7	257	2	182.0	364	− 33	112	42
Spain	55	23.6	1300	75	30.2	2265	36	28	74
All tropical nations	356	8.5	3017	498	11.2	5580	40	32	85
All temperate nations	1277	18.4	23,498	1564	23.6	36,866	24	28	57
World	1633	16.2	26,515	2062	20.6	42,446	26	27	60

Note: Data are given for developing countries planting over 5,000 hectares of tomatoes annually in either period and for representative developed countries. Countries that have more than 50 percent of their land area between the Tropic of Cancer and the Tropic of Capricorn are considered tropical.

Sources: FAO and Department of Agriculture and Forestry, Taiwan.

The yield of tomatoes in the tropics is much lower than in the temperate zones. In Table 5, for 1961-1965, the national yields in the tropics range from 1.7 to 15.8 t/ha, while in the temperate zone the range was 6.7 to 86 t/ha. By 1973-1977, the range was 1.8 to 22.9 t/ha in the tropics and 8.9 to 182 t/ha in the temperate zone.

Tropical region. Tomato production in the tropics has increased mainly from planting of new areas and secondarily from higher yields per hectare. A few countries, however, such as Taiwan, Mexico, and Brazil, have succeeded in increasing yields per hectare dramatically—by 8 t/ha or more. Most yields in the tropics are below 10 t/ha, and during the span of 12 years shown in Table 5 only three countries—Taiwan, Mexico, and the Philippines—were able to double their yields per hectare.

Of the tropical countries, those in South America have enjoyed the highest yields per hectare. All of them had yields over 10 t/ha, even in the early 1960s. In these countries, the major vegetable producing areas are in the highlands, which have generally cool and favorable weather for tomatoes.

Temperate region. In 1973-1977, three-fourths of the land planted to tomatoes worldwide was in the temperate zone, but this land area accounted for nearly 90 percent of world production. Most of the countries in the temperate region have a mild climate favorable to tomatoes. In addition, farmers in many of those countries utilize modern practices particularly suited to tomato production.

In a few affluent nations of the temperate region, yields exceeded 20 t/ha, even during 1961-1965. By 1973-1977, however, the yields in many developed countries had risen to over 30 t/ha. The Netherlands registered 182 t/ha in 1973-1977 and some other European nations, such as Belgium, Norway, and Denmark, obtained similar or higher levels. These extraordinary national yields, however, came from crops grown mainly or totally in greenhouses. Such high yields nevertheless demonstrate the potential of tomatoes under ideal conditions.

Product uses

Fresh market. Distinctions are made in some countries

between a fruit tomato and a cooking tomato, primarily on the basis of differences in quality characteristics. A fruit tomato is one that is eaten raw and served either as a fruit or vegetable. As a fruit it is eaten whole, like an apple, or cut into wedges and served as a refreshment or dessert. When used as a raw vegetable, it may be either sliced for sandwiches or cut into chunks for salads. In general, medium-to-large fruit with excellent taste, flavor, and color are preferred.

In contrast, the cooking tomato is generally baked, stewed, steamed, or made into a sauce for various foods. For such uses, size, shape, and color may not really matter and a more acidic tomato is preferred since the acidity blends with other ingredients during cooking.

Processing. The rapid development of the tomato-processing industry in developed countries in recent decades can be attributed to a series of interlinked activities. These include research and development, which led to the introduction of better varieties, more efficient techniques of production, and improved methods of processing. The ease and readiness with which tomatoes can be processed into various products that can be used for the manufacture of other foods make it a most popular agricultural product for food processors.

Most of the standards of quality for tomato products now in force in various countries are based largely on standards issued by the U.S. Department of Agriculture. In some countries, however, certain processing characteristics (such as maximum mold count, minimum sugar content, and freedom from artificial coloring) are governed by legislation. For this reason the quality standards required by importing countries should be taken into consideration when manufacturing tomato products for export.

The following are the most important characteristics of various tomato products.

• Whole peeled tomato (or canned tomato) is simply the product resulting from peeling, coring, and canning the tomato. Peeling and coring constitute at least 65 percent of the labor cost of processing. To reduce costs, less labor-intensive methods of peeling have been developed, including the use of steam, lye, light rays, and gas scalding. Coring, on the other hand, is done by hand and by machine. The methods selected are determined largely by the cost of labor, equipment, and the processing volume. If tomatoes do not meet the requirements for whole peeled tomato they may be canned diced, sliced, or as wedges.

•   Tomato pulp is the same as tomato puree and is made from raw tomatoes by separating the liquid and flesh portion from the seeds, cores, and other coarse or hard substances. Water is removed until the concentrated product contains from 8 to 24 percent salt-free tomato solids (from 4 to 6 percent natural solids).

•   Tomato paste is obtained by evaporating water from the pulped tomato to concentrate it further (from 24 percent to over 39 percent solids), with or without the addition of salt, spices, and chemicals. Today the terms "concentrate" and "paste" generally mean the same tomato product.

•   Tomato juice refers to the crushed, screened, and refined tomato pulp in an unconcentrated form, which contains finely divided insoluble solids from the flesh of the tomato. This product, usually intended for consumption without dilution or concentration, is unflavored except for the addition of a small quantity of salt. However, many new beverages that use tomato juice as a base include seasonings such as sugar, flavorings, citric acid, spices, and salt. Tomato juice also appears on the market in a concentrated form and contains from 20 to 24 percent natural tomato solids.

•   Catsup may be made directly from fresh juice after the removal of seeds, skins, and cores or it may be made from concentrated pulp. Processors generally prefer to use fresh juice because concentrated pulp loses some of its color during storage and requires additional processing. Sugar, vinegar, salt, onions, and spices are used in the manufacture of catsup. Hot sauce is made like catsup except that peeled and cored tomatoes with seeds are used instead of fresh juice or concentrated pulp. The other ingredients are the same but more sugar and onions are added, plus cayenne pepper to make the product spicier.

•   Tomato powder may be reconstituted as a drinking juice or an ingredient in soups. Dehydration is generally accomplished by roller and drum drying and by various spray-drying techniques. An ideal tomato powder should possess good keeping qualities and disperse readily in water, providing a product almost like a natural material in flavor, taste, color, and physical and chemical properties.

• Candy, chutney, and pickles can be made from tomatoes. Candied tomatoes have been produced in Taiwan on a commercial scale by slowly impregnating the fruit with syrup. Repeated boiling and soaking in syrups of progressively increasing sugar concentration raise the sugar concentration in the fruit until it is high enough to prevent spoilage. The tomato is washed and dried following impregnation and can be sold in this form or coated with a thin glaze of sugar or syrup.

• Tomato chutney is prepared by cutting the ripe fruit into slices and boiling them with vinegar until the fruit is soft. Spices such as ground chilies, garlic, ginger, and seasoning (salt and sugar) are added and the mixture is boiled again over low heat until the desired consistency is obtained.

• Green tomatoes are pickled in the same manner as cucumbers. Whole or sliced tomatoes are fermented in a brine solution, which must be stronger than that used for cucumbers to minimize gaseous fermentation. After complete fermentation, the tomatoes are soaked in water to remove the salt and then placed in distilled vinegar until a desired acidity is reached. At this point they may be prepared as sour or sweet pickles. Green tomatoes are usually mixed with cucumbers, onions, and cauliflower for the preparation of sweet mixed pickles.

Consumption patterns

Consumption of tomatoes is high in temperate countries—over 10 kilograms per capita per year—and lower in the tropical countries (Table 6). However, a clear upward trend in domestic consumption exists in tropical countries like Brazil, Ghana, and Indonesia as tomato production increases. In some countries, however, consumption is decreasing because of large exports. For example, Mexico exported 44 percent of its relatively large production in 1977, which resulted in a lower domestic consumption. If all tomatoes produced in Mexico that year had been used domestically, the per capita consumption would have been 132 kilograms instead of 6.8 kilograms.

In general, consumption patterns vary from country to

Table 6

Apparent average annual per capita tomato consumption* in several tropical and temperate countries, 1965 and 1977

| | Consumption | | |
	1965 (kg/capita)	1977 (kg/capita)	Increase or Decrease (%)
Tropical			
Brazil	7.0	11.1	58
Ghana	3 2	9.6	196
India	1.1	1.1	1
Indonesia	2.5	2.8	12
Mexico	9.2	6.8	- 26
Nigeria	4.0	3.7	- 6
Philippines	2.4	1.5	- 38
Sudan	9.1	8.8	- 2
Temperate			
Australia	14.1	12.7	- 10
Egypt	44.6	64.5	45
Greece	54.2	156.9	189
Italy	60.9	58.2	- 4
Japan	5.4	8.3	55
Morocco	14.0	18.6	32
U.S.	26.7	37.9	42
USSR	11.2	17.7	59

*Production plus net imports, divided by population.

Sources: FAO Production Yearbook and FAO Trade Yearbook.

country depending on the season and on income levels. For example, consumption in the Philippines has shifted from eggplant to tomatoes as incomes have increased. In the dry season, tomatoes rank as the most important vegetable, but they are second to eggplant during the rainy season when the tomato supply is limited. Similar patterns exist in other tropical countries.

Quality preferences

In Western countries the popularity of tomatoes is readily attributable to their attractive color, flavor, and versatility. Fresh tomatoes are sliced thin for sandwiches and salads and processed forms are used as ingredients in diverse foods from spaghetti to Bloody Mary cocktails. But in most tropical countries (especially in Asia, Africa, and some parts of Latin America) fresh tomatoes are cut into small pieces to be used as condiments in a variety of dishes.

Quality requirements for processing tomatoes are specific: high content of solids (at least 4.5 degrees Brix), low pH (about 4.4), firmness, easy peeling, crack resistance, and excellent red color.

In the United States a high solids content in the fruit has been a principal goal of most breeders of processing tomatoes in California. Maintaining high levels of solids, however, has been a problem because as yields increase the solids percentage decreases. A pH higher than 4.5 is undesirable because it increases problems with thermophilic organisms, making it difficult to get a safe product with normal processing techniques. Firm fruits reduce the amount of damage during transport from field to factory. Easy peeling reduces the time involved to remove the peel prior to processing. Crack resistance reduces fruit spoilage. Finally, deep red color is an excellent index of maturity, which in turn is directly related to the maximum flavor development of the processed product.

Individual preferences for fresh tomatoes vary considerably. For example, within the same family one person may prefer a red, ripe tomato, another a breaker (blossom end turning red), and still another a green tomato. In general, the Japanese and Chinese like low-acidity tomatoes because they eat them as fruit, but in most tropical countries, where tomatoes are used for cooking, high-acidity (more sour) tomatoes are preferred.

Large tomatoes are preferred for salads and sandwiches. Grading systems according to size have been established in developed countries, but in developing countries marketable size is usually not restricted within narrow limits. Hotels in developing countries, however, create a special demand for large tomatoes to serve to tourists.

Deep, round tomatoes seem to be preferred in developed countries but rural people in countries such as the Philippines, Ghana, and Ecuador are accustomed to flat, irregularly shaped ones.

Color preferences in tropical countries vary from country to country depending on season and use. For example, tomatoes are harvested in Taiwan in the mature green stage and sold in the market before completely ripening. A dark green shoulder with the red color progressing from blossom end to stem end is most desirable. In a local survey of consumption patterns, Taipei consumers preferred red color, but in Kaohsiung more people favored green fruit.

Tomatoes with green shoulders are preferred in Brazil and Colombia because of uneven ripening. At a given time, some fruits are red ripe and ready for use while others are still turning color. Similarly, Mexican and U.S. tomatoes in Florida are harvested mature green for shipping to long-distance markets.

Europeans and Americans generally prefer red, ripe tomatoes, but there are exceptions. In the United States most northern Ohio greenhouse growers produce pink-fruited tomatoes, whereas red-fruited varieties are grown in southern Ohio. In Japan and Korea pink tomatoes are generally preferred to red ones.

Tomatoes that have several thick and comparatively meaty

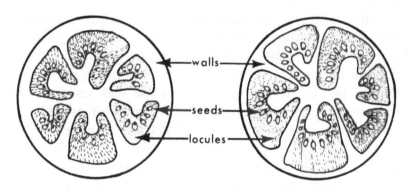

Figure 7. Cross section of multilocular tomatoes. Thick walls are more desirable than thin walls.

central areas are generally considered better than those with several locules (seed cavities) but less central flesh and larger locules (Figure 7). The fibrous core should be small. In addition, other fruit qualities such as firmness and small blossom-end closure and stem-end scar contribute to attractiveness and market appeal.

Nutritional composition

Tomatoes are not particularly nutritious (Table 7), but they can be a major source of minerals and vitamins if consumption is encouraged. Tomatoes do not rank high in the concentration of any particular dietary components. For example, among the main fruits and vegetables eaten in the United States, tomatoes rank sixteenth as a possible source of vitamin A and thirteenth as a possible source of vitamin C. Nevertheless, because of the

Table 7
Nutritive values of 100–gram edible portion of raw and processed tomatoes

Nutrient	Raw	Canned*	Catsup	Juice
Water (%)	94	94	69	94
Food energy (cal)	19	21	106	19
Protein (g)	0.7	0.8	1.8	0.8
Fat (g)	trace	trace	0.4	trace
Carbohydrate (g)	4	4	25	4
Calcium (mg)	12	6**	22	7
Phosphorus (mg)	24	19	50	18
Iron (mg)	0.4	0.5	0.8	0.9
Potassium (mg)	222	217	363	227
Vitamin A value (I.U.)	822	900	1399	798
Thiamin (mg)	0.05	0.05	0.09	0.05
Riboflavin (mg)	0.04	0.03	0.07	0.03
Niacin (mg)	0.7	0.7	1.6	0.8
Ascorbic acid (mg)	21	17	15	16

*Solids and liquids.
**Applies to product without added calcium salts.

Source: USDA Home and Garden Bulletin, No. 72.

large consumption of tomatoes in the United States, they rank third as the actual source of both vitamins. In developing countries the consumption of tomatoes can be increased, causing them to become a larger source of vitamins, if greater supplies are more readily available to consumers all year at relatively inexpensive prices.

5
Varieties, Seed Production,
and Distribution

Unadapted varieties

Scientists from the tropical regions of Africa, Asia, and Latin America have identified the reasons tomato varieties grown in their areas are unadapted (Table 8). The reasons fall into three main categories: susceptibility to diseases, low fruit-setting ability, and poor quality fruits for fresh market or processing.

Susceptibility to diseases. Bacterial wilt (*Psuedomonas solanacearum*) may be the most important tomato disease in the humid lowland areas. It is caused by a soil pathogen that thrives at relatively high temperatures. Yield losses can range from 30 to 100 percent. A susceptible variety grown in tropical soils nearly always becomes infected.

Several species of nematodes attack vegetables, but the most extensively studied are the root-knot nematodes (*Meloidogyne* spp.). They damage the root system through the formation of knots and galls and cause rotting, thereby reducing growth and yield. They also increase the severity of fusarium, verticillium, and the bacterial wilts of tomatoes.

Fusarium and verticillium wilts—soil-borne fungal diseases of tomatoes—are rare in the lowland tropics but can be serious problems in the highland areas. The fungus that causes fusarium wilt affects only the cultivated tomato and a few wild tomato species. Verticillium wilt, however, attacks many plants such as potato, eggplant, pepper, and okra. Both pathogens generally enter through the roots and pass upward into the xylem. They cause various degrees of wilting resulting in low productivity of the crop.

Table 8
Major problems of temperate-bred tomato varieties grown in the tropics

	Africa	Asia-Pacific	Latin America
Susceptibility to:			
Bacterial wilt (*Pseudomonas solanacearum*)	•	•	•
Early blight (*Alternaria solani*)	•	•	•
Late blight (*Phytophthora infestans*)	•	•	•
Viruses	•	•	•
Fusarium wilt (*Fusarium oxysporum*)	•	•	•
Root-knot nematodes (*Meloidogyne* spp.)	•	•	•
Septoria leaf spot (*Septoria lycopersici*)		•	•
Gray leaf spot (*Stemphylium solani*)		•	•
Leaf molds (*Cladosporium fulvum*)		•	
Black leaf molds (*Cercospora fuligena*)		•	
Southern blight (*Sclerotium rolfsii*)		•	
Bacterial canker (*Corynebacterium michiganense*)			•
Bacterial spot (*Xanthomonas vesicatoria*)			•
Poor fruit set due to:			
Heavy rainfall	•	•	•
High temperature	•	•	•
Other:			
Poor quality fruits	•	•	•
Acid soils		•	
Spider mites			•

The most important leaf diseases in the lowland tropics are leaf molds, gray leaf spot, viruses, septoria leaf spot, black leaf molds, southern blight, and early blight. The degree of yield losses caused by these diseases varies from country to country in the tropics. In India, for example, leaf curl virus is the most devastating, but in the Philippines leaf molds are more serious.

Late blight, a fairly common disease of tomatoes grown in the highlands, can also be severe during the winter season in some lowland areas. When the nights are cool and the days only moderately warm with abundant moisture the pathogen multiplies rapidly and causes heavy damage. Also, bacterial canker, bacterial spot, spotted wilt virus, and yellow top viruses can be very serious problems of tomatoes in the highlands, as in Brazil and Colombia.

Low fruit-setting ability. For optimum fruit-setting, tomatoes require nighttime temperatures of 15° to 20° C. Unfortunately, the minimum temperature in the lowlands of most tropical countries rarely drops to 20° C. even during the cooler months. In addition, many areas in the tropics have extremely high daytime temperatures, which are unfavorable for normal development of pollen grains and fruits. To be grown in the lowlands in the summer, tomato varieties often must be moisture tolerant as well as heat tolerant, because in many locations the period of high temperatures coincides with the period of high rainfall.

Poor quality. The quality of tomatoes sold in tropical markets is generally poor. Usually tomatoes are harvested during the summer before they attain the mature green stage, resulting in poor color and flavor development. Even when growing conditions are favorable, most tomatoes reach the market over-ripe, unevenly red, and blemished. Moreover, tomatoes for the processing industry often do not have the attributes—such as deep red color, high solids content, appropriate pH, and proper viscosity—that are needed to produce superior canned tomatoes. Since few tomato-processing operations exist in tropical countries, there has been little reason for tropical breeders to select varieties that combine these attributes.

Susceptibility to pests. Sometimes heavy infestations of spider mites and tomato fruitworm attack tomatoes, however their impact on yield is less serious than that of diseases. Researchers, therefore, give less attention to these problems than to disease problems. The most dangerous insect enemies of tomatoes are those that transmit virus diseases—aphids (*Myzus persicae*) for leaf roll, whitefly (*Bemesia tabaci*) for leaf curl, and thrips (*Thrips tabaci*) for spotted wilt virus.

Current advances in tomato breeding

Recently, researchers have made remarkable gains developing disease-resistant varieties and breeding lines, understanding the basic causes of low fruit-setting, improving quality of both processing and fresh-market tomatoes, and transferring useful traits from wild species to cultivated varieties.

Disease resistance. Significant advances in the development of disease-resistant varieties have resulted from the collective efforts of many scientists to find resistance genes in wild species of tomatoes and to transfer these genes to adapted but susceptible local varieties. Tomato breeders throughout the world share results and freely exchange seeds of potential parents. Several genes for resistance to specific diseases have been successfully bred into tomato varieties that are commerically grown in many temperate countries. For example, resistance to fusarium and bacterial wilt were derived from *Lycopersicon pimpinellifolium*, resistance to TMV (tobacco mosaic virus) from *L. peruvianum*, and resistance to early blight from *L. peruvianum, L. hirsutum,* and *L. pimpinellifolium.*

Understanding how the different resistance genes are inherited is of great importance to plant breeders. It allows them to determine the appropriate breeding procedure for transferring a specific gene to a horticulturally acceptable variety. Table 9 shows the mode of inheritance of some important disorders. Genes for resistance to these disorders have been used also in tropical areas such as Puerto Rico, Malaysia, the Philippines, and the West Indies, where tomato improvement had been done on a piecemeal basis mainly because of the lack of funds. Research results in many tropical countries come in a trickle from university professors who squeeze their already limited budgets for instruction to conduct some research. Many of their findings have given a good start to tomato improvement in the tropics.

AVRDC has the largest tomato-breeding program in the tropics. An interdisciplinary team of scientists has been able to combine bacterial wilt resistance with good fruit-setting ability in several breeding lines that have been evaluated in tropical environments in cooperation with local scientists. In tests in many tropical areas of Asia and the Pacific, these breeding lines have usually performed better than local varieties because of their bacterial-wilt resistance and heat tolerance. These two traits are essential if tomato varieties are to grow successfully in the tropics.

Table 9
Common tomato disorders in the tropics and their mode of inheritance

Disorders	Mode of inheritance
Bacterial wilt	complex
Bacterial canker	complex
Bacterial spot	complex
Early blight	
collar-rot phase	single incompletely dominant gene
leaf spot phase	two or more recessive genes
Gray leaf spot	complex
Late blight	single dominant gene and complex
Leaf molds	single dominant gene, pathogen is very mutable
Septoria leaf spot	single dominant gene plus one or more modifiers
Fusarium wilt race 1	single dominant gene
Fusarium wilt race 2	single dominant gene
Verticillium wilt	single dominant gene
Spotted wilt virus	single dominant gene
Tobacco etch virus	single recessive gene
Tobacco mosaic virus	single dominant gene
Root-knot nematode	single dominant gene
Heat tolerance	complex
Fruit cracking	complex
Blotchy ripening	complex

Note: Seed samples of resistant accessions or breeding lines may be requested from AVRDC, the U.S. Department of Agriculture Regional Plant Introduction Center at Ames, Iowa, and various experiment stations in the United States.

The experience in Malaysia is a case in point. The Malaysian Agricultural Research and Development Institute began to receive AVRDC materials in 1974, and they were evaluated for four consecutive seasons in the same field. Ordinarily this is an undesirable practice, but it was done deliberately to build up the population of pathogens. The AVRDC materials yielded 14 to 28 t/ha, while the check varieties were completely destroyed by bacterial wilt. This was considered a breakthrough by Malaysian scientists. Previously tomatoes could be grown at the station only if they were grafted to wilt-resistant eggplant root stock.

Currently researchers at AVRDC are incorporating resistance to several other diseases into breeding lines. Some lines now carry resistance to leaf molds, gray leaf spot, and root-knot nematodes.

Acceptance of the superior lines varies from country to country. Today, only a few tropical countries have extension programs in vegetable crops that are capable of transferring this technology to farmers. Some national programs—including those in India, the Philippines, Malaysia, Papua New Guinea, and Thailand—have begun using AVRDC materials for tomato improvement.

Fruit-setting. Varieties with the ability to set fruit under high temperatures and excessively wet conditions are critical for tomato production in the tropics. The ability to set fruit under high temperatures is a characteristic that can be bred into a plant just like other heritable, desirable traits. In the early 1940s, F. W. Went discovered that with commercial tomato varieties, fruit-set is possible over a limited range of night temperatures (about 15° to 20° C.). Experiences at AVRDC have shown that under natural conditions tomatoes exposed to high night temperatures are also subjected to high day temperatures, suggesting that both could be important to fruit-setting. In fact, if the plant is subjected to extremely high temperatures (40° C. or higher), unfruitfulness will persist for a week or more due to physical destruction of the pollen grains.

At AVRDC, scientists have screened more than 4600 tomato accessions and have found only 39 that have heavy fruit-setting ability under high temperatures. The physiological basis of the heat tolerance of these lines varies. Some have high pollen viability, some have high ovule viability, some possess high pollen production and lack stigma exsertion. Several of these accessions have been crossed to combine their attributes into a single genotype.

Excessive moisture also affects the fruit-setting ability of tomatoes. In one study at AVRDC, conducted during the cool-dry period to nullify the effect of temperature on fruit-setting, the response to moisture of Nagcarlan (heat-tolerant), LA1421 (moisture-tolerant), and White Skin (neither moisture-tolerant nor heat-tolerant) was also investigated. Simulated rain (from sprinklers) reduced the fruit-setting of all three varieties, as well as the dry matter content of their roots and tops. The reduction was magnified when the drainage was poor. But the root growth and fruit-setting rates of Nagcarlan and LA1421 were less seriously affected by excess water in the air or poor drainage than those of White Skin. This study demonstrated the need to have a separate breeding program for hot-dry and hot-wet conditions in the tropics. For hot-dry conditions it is sufficient to screen materials under high temperatures. For hot-wet conditions, however, breeding lines should be screened under high temperatures and excessive moisture (both in the air and in the soil), which are typical of the rainy season in most tropical countries.

Improved quality. Plant breeders have succeeded in creating fresh-market tomatoes that have a smooth, deep, round appearance in contrast with the flattened and segmented appearance typical of old cultivated varieties. Fruit color has been changed from pale red to a more uniform deep red. And taste and flavor have been improved, though both depend on when the tomatoes are harvested. Vitamin C and sugar content decrease when tomatoes are harvested at the mature green stage and ripened in storage or while in transit. In vine-ripened fruits the opposite occurs. Since the amount of sugars, acids, and volatile compounds determine the flavor, tomatoes harvested red-ripe taste better than tomatoes of the same variety harvested at the mature green stage. Today, varieties of varying color, acidity, and shape have been bred and are grown commercially in many parts of the world.

Three mutants have been discovered that may lengthen the shelf life of tomatoes, particularly those grown in the tropics. Mature fruit with the *rin* (ripening inhibitor) gene remain firm and in good condition for several months after harvest. The *nor* (nonripening) gene affects carotene synthesis and fruit softening. Sound fruits with this gene remain firm and green for a long time. The *nr* (never ripe) gene permits fruits to turn red at normal times, but they develop pigmentation slowly and never become deep red regardless of how long they stay on the plant or in storage. Fruits of this mutant retain the texture and low sugar content of mature green fruits.

Many plant breeders have contributed to the development of varieties with good processing qualities, such as high content of soluble solids, a pH of less than 4.5, high content of alcohol-insoluble solids, and deep red color. In addition, firm-fruited varieties have been developed that can remain on the vine in good condition for several weeks after the fruits reach full color. Moreover, varieties with a shorter core—a desirable trait for canned whole peeled tomatoes—have been bred.

Other useful traits. In addition to disease resistance and quality, many other desirable traits are being bred into culti-vated tomato varieties, such as the linkage of the anthocyanine-less gene to TMV resistance, determinate growth characteristics, the jointless trait, and uniform ripening.

The linkage of the anthocyanineless gene (*ah*) to TMV resistance has a practical application. A normal seedling has a purple stem whereas the single recessive *ah* gene imparts a green stem. This means that when a purple-stemmed plant is crossed with a green-stemmed plant, all the first generation (F_1) seed-lings have a purple stem. In the second generation (F_2), three-fourths of the seedlings have purple stems, the remaining one-fourth have green ones. The latter breed true to type. If linkage is complete and the gene is in the background of a TMV-resistant parent, selecting for green stems is an easy indirect way to select for resistance. Another gene, netted virescent (*nv*), is also linked to TMV resistance and has the same mode of inheritance. One drawback of this technique, however, is that varieties that have either the *ah* or the *nv* gene are weaker during the seedling stage than their normal counterparts.

Determinate growth characteristic (Figure 8) is controlled by a single recessive gene, *sp*. This trait is necessary, along with compact vine and concentrated fruit-set, in machine-harvested varieties. Determinate growth is also convenient for tomatoes intercropped with sugarcane, rice, or maize. On the other hand, sprawling, unlimited indeterminate growth has made greenhouse production in Western Europe a continuing, productive enter-prise. Indeterminate types are normally trellised and cultivated more intensively than determinate ones.

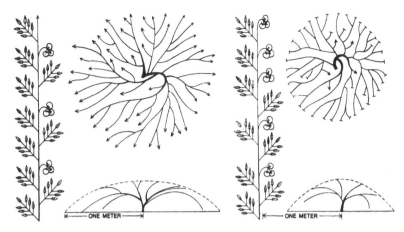

Figure 8. Growth characteristics of indeterminate tomatoes, left, and determinate tomatoes. (From "The Tomato" by C. M. Rick. Copyright © 1980 by *Scientific American*, Inc. All rights reserved.)

Stalk (pedicel) retention on the fruits is undesirable in machine-harvested tomatoes because the stalks can damage other fruits in a load and they affect the flavor of the processed product, especially if the load is not processed immediately. Therefore, processors prefer varieties with low stalk retention. A jointless gene ($j2$) reduces stalk retention because the fruit is detached directly from the stalk instead of the stalk's being separated from the fruit cluster. Although this gene has some yield disadvantage, it is being used in several new varieties because of its potential in reducing shattering and fruit damage.

Uniform ripening is imparted by the gene u, which eliminates the dark green shoulder of unripe fruit. In many countries where uniform ripening is desirable, the gene is bred into cultivated varieties. But in such countries as Taiwan, Colombia, and Brazil, tomatoes are bred with the nonuniform ripening gene (U).

Other mutant genes are discovered from time to time by tomato geneticists around the world. New genes are described, mapped, and reported in a number of journals. One important outlet is the informal *Report of the Tomato Genetics Cooperative* published in the United States (available from C. M. Rick; Davis, California—see Appendix B).

No single tomato variety carries all known desirable attributes for cultivation in the tropics. Yet, in extensive trials conducted in the tropical areas of Africa, Asia, Latin America, and the Pacific by AVRDC scientists, improved lines have clearly demonstrated superiority over local varieties. In addition, in the United States materials being developed under hot-dry conditions in Texas and under wet conditions in Florida generally do well in tropical areas that have similar growing conditions, provided bacterial wilt is not a problem. Varieties from Hawaii give equally encouraging results. Scientists in the tropics who plan to evaluate varieties or conduct a breeding program may want to request materials from AVRDC and experiment stations in those states.

Seed production

No input in the production of crops gives greater results with less effort than good seed. In fact, no amount of fertilization, pesticides, or good cultural practices will give profits if poor seeds of poorly adapted tomato varieties are planted. Seeds that are undamaged, have good germinating ability, and are free from mixture with other varieties and from seed-borne diseases can be supplied only with appropriate seed production technology. It would surely be a big advantage for any developing nation to produce its own tomato seeds.

Production of tomatoes for seeds requires experienced seedsmen. They should have a good background in genetics and plant breeding to ensure genetic purity and high quality and they should know the technology of seed preservation and methods of seed packaging to maintain good germination. Skilled manual laborers, who cannot be replaced by machinery, and highly intensive cultivation are also essential.

Domestic requirements. In 1977, world seed requirements for transplanted tomatoes (common varieties) added up to 439,400 kilograms worth US$21 million (Table 10). If developing countries had produced their own seeds, they could have saved millions in foreign exchange.

Common varieties versus hybrids. Since the 1960s, hybrid seeds have been used in greenhouses in Japan, the United States, and several Western European countries. Hybrids now account for 100 percent of Japan's fresh-market tomatoes, and their

Table 10
World requirements for seed of common tomato varieties, 1977

	Planted area* (thou. ha)	Seed required** (tons)	Value† (thou. $ U.S.)
World	2,197	439	21,090
Developed	580	116	5,570
North America	221	44	2,120
Western Europe	309	62	2,970
Oceania	9	2	90
Other	41	8	390
Developing	925	185	8,880
Africa	138	28	1,320
Latin America	215	43	2,060
Near East	383	77	3,680
Far East	188	38	1,800
Other	1	0.2	10
Centrally planned	692	138	6,640
Asian	261	52	2,510
Europe, USSR	431	86	4,140

Source: FAO Monthly Bulletin of Statistics

**Computed at 200 g/ha, the optimum rate for transplanted tomatoes
†For common varieties, based on $48/kg—the average price of tomato seeds taken from several U.S. and Asian seed catalogs

popularity is growing in the United States for commercial and home garden uses. About 20 percent of the tomatoes used for processing in California are grown from hybrid seeds. Plantings in other developed countries have also increased tremendously. In developing countries, however, research workers and growers are still attempting to determine whether hybrid tomatoes are really superior to common varieties.

Seed companies claim certain advantages for hybrids over common varieties (standard varieties) such as better quality, higher productivity, better resistance to diseases, vigorous growth, better adaptability, and earlier maturity. However, common varieties now exist that have traits comparable to hybrids. Nevertheless, in countries such as Japan, the Netherlands, Belgium, Norway, and Denmark, where the seed trade is independent of government control, common varieties have disappeared from seed catalogs for several reasons. The huge investment involved in the developing of common varieties is lost within a few years. If a company develops a common variety, a farmer or a competing company can buy the seeds one year and produce seeds of the variety the following year. This does not happen with hybrid seeds because the parental lines are kept secret. Another reason for the rapid adoption of hybrids in developed countries is the extensive promotional campaigns of seed companies.

Government institutions in developing countries are obliged to produce common varieties rather than hybrids because they need tomato seeds that farmers can reproduce themselves. Private seed companies engaged in production, processing, packaging, storing, marketing, and distribution exist in only a few developing countries. As long as government takes care of seed production and no private seed companies operate, the use of common varieties will continue in most developing countries.

Hybrids are generally more expensive than common varieties, principally because hybrids are hand-pollinated whereas common varieties are generally planted in isolation and allowed to pollinate themselves to produce true-breeding seeds. The average price of hybrid seeds in various seed catalogs is 4 to 15 times the price of common varieties. In general, however, seed amounts to only 2 to 4 percent of the total cost of tomato production, so farmers in developed countries are quite willing to use hybrids.

Policymakers can influence the adoption of varieties. In the summer of 1978, local authorities in Taiwan required farmers to plant Known You 4 (a hybrid) or White Skin if the tomatoes were being produced for export. Consequently, about 200 hectares (two-thirds of the total area devoted to export tomatoes) were planted to these varieties in one season. Production programs could also encourage the use of a specific variety if loans are given as inputs (seeds of a specific variety, fertilizers, and pesticides).

Production of hybrid tomato seeds is labor-intensive. One hectare of tomato seed production requires 50 skilled pollinators for 30 days, or the equivalent of 12,000 hours of labor. Even the use of male-sterile lines in commercial production of hybrid seeds cuts the labor need by only 50 percent because pollination still has to be done by hand. The experience in Asia illustrates this.

Until the early 1960s, Japan was the main hybrid tomato seed producer for a number of U.S. and European seed companies. By the 1970s, however, seed production had moved to Taiwan, and in 1977, 15,000 kilograms of hybrid tomato seeds, or enough to plant 100,000 hectares, were produced there. The change in location was brought about primarily by a shortage of pollinators caused by industrial development in Japan's rural areas and by income differences between agriculture and other industries. Higher agricultural wages were demanded, increasing the cost of producing seeds. There was also a decrease in the number of seed-growing farms because of expansion of industrial and residential areas. Similar changes are occurring in Taiwan now. Whether the country will continue to be suitable for hybrid tomato production is an open question because of rising labor costs.

Developing countries with inexpensive labor have the potential to produce hybrid tomato seeds if other factors are favorable. In fact, countries such as Mexico, Chile, Costa Rica, Guatemala, and India have already taken a minor share of the hybrid tomato seed trade. Trial plantings have also been made in the Philippines, Thailand, and Indonesia. These agriculturally based countries are planning to expand seed production programs.

Production of hybrids

There are four essentials for starting a commercial hybrid tomato seed operation: capital, technology, inexpensive labor, and appropriate climatic conditions. Both capital and technology can be transferred from developed countries. The availability of low-cost labor is a temporary condition depending on the economic growth of a country. Climatic conditions, of course, are more permanent and cannot be modified without large capital expense.

Not all countries with low-cost labor will be able to meet the strict conditions required to produce high-quality hybrid seeds. The farmers must be industrious and skilled in applying intensive methods of tomato seed growing. Pollinators must have sharp eyes to spot all flowers and nimble fingers to minimize damage to the flowers. There are no holidays in this type of production because for a single F_1 combination emasculations and pollinations must go on at least 8 hours a day for 30 days for each cropping season.

Hybrids, like other tomatoes, need long days with plenty of sunshine. Fruit-setting should occur during periods with little or no rainfall and cool night temperatures (15° to 20° C.) for optimum fruit-setting and seed development. A dry period at harvest permits efficient seed processing and the use of sunshine for drying the seeds.

Contractual agreements. Commonly a local seed company accepts seed orders from a foreign company. The local company, in turn, orders the seeds from a grower. Thus, two contracts are made: between the local company and the foreign company and between the local company and the grower. In both contracts, certain conditions are spelled out in connection with the growing, pollination, and isolation requirements; harvesting; curing, separating, and cleaning the seeds; germination and purity of the seeds; and secrecy on the part of the grower to keep stock seeds from falling into the wrong hands. In addition, the contract specifies the exact area to be planted, contract price, and mode of payments and includes a statement to the effect that both contractors (local and foreign) reserve the right to enter the grower's area, at any time and at their expense, to examine, reject, and remove any plants that in their opinion are offtypes.

Finding growers. To initiate seed production, prospective growers must be found who are willing to learn the various operations for producing high-quality hybrid seeds. A thorough explanation and practical demonstration of the harmful consequences of cross pollination and seed mixture should be presented to growers. In addition, a crew of experienced pollinators from the local company should be assigned to each new grower for at least two weeks, so the grower and his hired pollinators can observe the discipline, dedication, and skillfulness of experienced pollinators.

Field inspection. To produce 98 percent purity in seed lots of hybrid seeds, a technician who represents the seed company regularly inspects the fields early in the growing season and during emasculation, pollination, harvesting, and seed processing. One technician can handle 8 to 10 hectares, depending on the distance between growing areas and on whether or not plantings are staggered. In addition to the company representative, experienced growers usually conduct a daily inspection, too. Close attention is paid during the emasculation and pollination period, which lasts for about a month.

Emasculation and pollination. Several methods of emasculation and pollination exist, but the most efficient—in terms of seed yield and labor—is the one seed growers in Taiwan use. Emasculation of the female parent begins when the second flower cluster reaches the bud stage. In the afternoon, anther cones are gently removed with forceps. Emasculation continues as other flower clusters reach the bud stage—through the sixth or seventh flower cluster. Part or all of the petal is kept to serve as an indicator of stigma receptivity. When the petal of the emasculated flower starts to turn bright yellow, the stigma is ready for pollination. All flowers that escaped emasculation are removed as soon as possible to avoid any contamination.

In pollination, the usual ratio of male to female parents is 1:6. Male parents are generally sown one to two weeks earlier than the female parents to provide sufficient pollen when the females are ready for crossing. They are planted in separate fields. Only the female parents are pruned and staked. Pollen is collected from the male parent by shaking the flower either by hand or with a vibrator. The collected pollen is massed in a wide-mouthed container where the index finger can be dipped to reach it. The index finger then lightly touches a receptive stigma to accomplish pollination. After pollination, two or three sepals are removed to serve as a marker for a hand-pollinated fruit. This is especially useful at harvest time.

Harvesting, seed extraction, and drying. Only fully red, ripe fruits that are hand-pollinated as shown by the marked fruits (two to three sepals removed) are harvested. Care should be given to discard unpollinated fruits because one bad fruit could contaminate the whole lot of seeds and cause the seed company to reject the lot.

Sometimes the tomatoes are stored in a cool place for a few days before seed extraction, especially if large quantities are harvested in one day. Usually, however, the fruits are crushed in the field and sieved to separate the skin and most of the pulp from the seeds. Only the seeds with juice have to be transported to the grower's house. There the seeds are fermented for one or two days to separate the seeds from the mucilagenous materials attached to them. The juice with seeds must be stirred thoroughly three to four times a day to encourage uniform fermentation. On the second or third day, the fermented mixture is placed in a fine sieve and rinsed with running water. The water is drained by putting the seeds into cloth bags and squeezing the water out by hand. Some farmers put the bags in a machine spinner to remove the water. Then the seeds are spread out to dry in the sun for one or two days. When the seeds register about 8 percent moisture content, they are ready for delivery to the local seed company.

Seed delivery and payment. Even before the seeds of a cross are delivered, a company representative has taken samples at two or three different times from the grower's field for germination and purity tests. If the local company has sufficient cash, the grower gets paid the contract price as soon as delivery is made. The mode of payment, however, varies from company to company and from one contract agreement to another. For instance, some companies pay 70 percent of the amount within one week and the remaining 30 percent after the results of the standard germination and purity tests are known. Some companies delay payment until they obtain full payment from a foreign company. Mode of payment is a critical aspect when companies are competing to recruit growers.

Seed distribution

Finding vegetable seeds that will germinate and grow with vigor is a problem for farmers and home gardeners in many tropical countries. Most vegetable seeds are imported from countries in the temperate zone. They arrive in large vacuum-sealed tin cans and then are repacked in paper containers or put in jars for retailing. When the dry seeds taken from the vacuum-sealed cans are exposed to the hot, humid conditions of the tropics, they absorb moisture and deteriorate rapidly. Tomato seed treated this way will increase from about 5 percent moisture content to 9 percent within two months. In six months, germination will be reduced from about 93 percent to 70 percent, and vigor will deteriorate from 85 percent to 60 percent. More sensitive vegetable seeds are worthless within one month after being removed from the cans.

Studies in the Philippines, however, have demonstrated that if the tin can is opened and the seeds repacked in an air-conditioned and dehumidified room (10° to 15° C. and 50 to 60 percent relative humidity), germination and vigor of seeds can be maintained several months longer, even if the seeds are subsequently kept under ordinary room temperatures. The seeds should be repacked in moisture-resistant containers made from polyethylene or, better, cellophane-aluminum-polyethylene.

Vegetable seeds are usually distributed through private seed companies, government institutions, or both.

Government institutions. In most developing countries seed production and distribution is left to government experiment stations or colleges of agriculture. If any commercial seed companies exist, they function primarily as import agencies.

Because of limited staff and facilities, government institutions have difficulty meeting the demand for vegetable seeds. To make matters worse, government agencies often attempt to introduce too many kinds of vegetables and evaluate them for yield and adaptability under limited environmental conditions.

Usually seeds of tomato varieties that show superior performance are mass-produced and either distributed free or sold at cost to farmers and home gardeners. Because such varieties lack thorough testing they do well only in a few locations and are generally low yielding elsewhere. Thus, farmers and home gardeners become disappointed and return to old varieties, which are similarly low yielding but more dependable. And since government institutions compete with the few individuals and companies who would like to be involved in seed production, the development of a private seed industry is retarded.

Private seed companies. Seed production and distribution in the Philippines represents a typical system in a developing country (Figure 9). In most developing countries, there is no private seed grower or company engaged in production, processing, packaging, storing, marketing, and distribution on a large scale. Seed companies in the Philippines are actually seed importers and they distribute seeds through dealers and sometimes through government stations. In fact, the Bureau of Plant Industry assumes a part of the responsibility of producing and distributing seeds. Some seed dealers go into seed production of tomatoes and native vegetables themselves or contract for small quantities of seed from local vegetable growers.

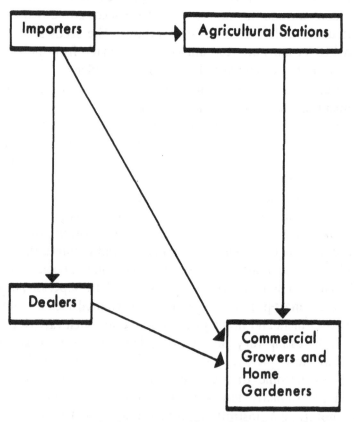

Figure 9. Organization of vegetable seed distribution in the Philippines.

In developed countries, a grower usually has only one source of tomato seeds—private companies. In Japan, for example, seed companies are quite independent of government experiment stations and universities. Takii Seed Company, the largest and most developed, maintains its own research staff. The company is equipped with modern facilities for processing, drying, and testing seeds. It is run by an experienced team of specialists in various fields, including vegetable breeding and marketing.

Seeds for commercial growing are produced through contracted local seed farms, and production is supervised by the company. In addition, the company sponsors an annual meeting to resolve problems and discuss trends in seed production, processing, and storage. These efforts ensure high yield of good

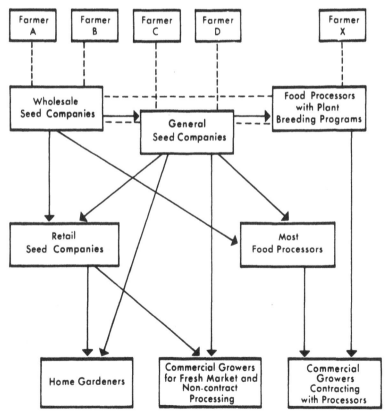

Figure 10. Organization of vegetable seed production and distribution in the United States. (Source: H. M. Munger, Cornell University.)

quality and genetically pure seeds that the company sells to dealers and growers.

In some cases, seeds are produced through contracted foreign seed companies. Taiwan seed companies, for example, have been producing hybrids and varieties for seed companies in Japan. The arrangement works well because they produce high quality seeds and labor is relatively less expensive in Taiwan than in Japan.

Companies are also the main source of vegetable seeds in the United States (Figure 10). They are well financed, well organized, and adequately staffed. There are numerous types of companies—such as wholesale seed companies, general seed companies, and food processors—that maintain stock seeds of established varieties, breed new ones, multiply these seeds in quantity, and sell them directly or indirectly to growers. These companies either produce their own seeds or contract with seed growers or other companies to produce seeds for them. There are also organizations such as retail seed companies and food processors that identify and purchase suitable seed lots of nearly all varieties they supply. They also sometimes maintain stock seeds.

6
Cultural Practices

Tomato cultivation can be classified by whether the plants are grown outdoors or under protection, whether the plants are grown with supports or on the ground, and whether the plants are transplanted or direct-seeded. Most of the world's tomatoes come from transplanted, unsupported plants grown outdoors.

Cultivation outdoors or under protection

Usually tomatoes are grown in open fields. This method, called outdoor cultivation, is widely used by home gardeners, small commercial growers, and large-scale farmers. Cultural practices vary from primitive to highly sophisticated.

Protected cultivation takes many forms ranging from mere rain protection to complete environmental control. With sufficient investment, the following factors can usually be adequately controlled: weather elements (rain, snow, and wind), light, temperature, atmospheric composition, and root environment. (The use of hydroponic culture is an example of modifying root environment: the tomato roots are anchored in sand, gravel, or some other inert medium, and all nutrients are supplied in water solution.)

In tropical areas such as the Philippines and Colombia, some commercial growers raise tomatoes under structures covered with plastic film to protect the plants from heavy rains. Usually the sides of the structures are left open for ventilation and illumination. The roofing lowers the incidence of insects and diseases and makes pest control easier because rain cannot wash

Greenhouses covered with plastic film are used by commercial growers in Colombia to protect tomatoes from summer rains. The greenhouse has open sides for ventilation and illumination. (Source: Instituto Colombiano Agropecuario.)

away applied chemicals. The roofing also reduces leaching of fertilizer. In highland areas, the warming effect of the plastic roofing is beneficial to tomato growth, but the reduction in light intensity is not. In lowland areas, the increased temperatures can be a serious problem.

In Taiwan some growers enclose tomato plants in nylon nets. The nets decrease wind speed by half; break the force of raindrops, which slows soil erosion and prevents mechanical damage to the leaves and flowers; and reduce insect damage. Nets are not practical in locations that have many cloudy or foggy days because they lower light intensity by at least 30 percent. Moreover, if the nets are not opened from time to time, poor air circulation may raise incidence of disease.

In Israel, Japan, and South Korea, temporary structures called high tunnels, made of wood, bamboo, or metal covered with plastic, have become popular. They cost less than permanent greenhouses and, because they are usually not heated, operating expenses are low. In Japan, prefabricated tunnels can be rented from private companies that erect and dismantle the structures.

Low tunnels are another type of protection used in Korea. Farmers establish tomatoes in the field during late winter or early spring and cover them with low tunnels constructed of wire or bamboo supports covered with plastic. When the danger of frost is past, the tunnels are removed to permit normal field cultivation. Many Taiwanese farmers also use nylon and plastic tunnels in the production of transplants.

Production in permanent greenhouses is a highly technical, intensive, and expensive agricultural enterprise, but it can be very profitable if done properly. Greenhouses may be covered with glass, pliable plastic film, rigid plastic, or fiberglass. The development of plastic film and fiberglass has lowered construction costs in recent years. As a result, production of greenhouse tomatoes has grown rapidly in many countries, especially in the less affluent ones.

Protected cultivation permits tomatoes to be produced in temperate countries during the winter because the plants' growth and development can be precisely controlled through regulation of soil moisture, soil fertility, light, temperature, relative humidity, and carbon dioxide concentration. In heated greenhouses tomatoes are protected from low temperatures and snow. Greenhouses have long been used for food production in Western Europe, especially for tomatoes, lettuce, and cucumbers. In addition, tomato seedlings for summer outdoor cultivation are usually started in greenhouses from late winter to early spring or in structures such as hotbeds or cold frames.

Highly advanced, environmentally controlled greenhouses exist in certain oil-rich nations. Abu Dhabi has an experimental integrated system developed by the University of Arizona Environmental Research Laboratory that can provide power, water, and food on desert coasts. Waste heat from an engine-driven electric generator is used to desalt water. The fresh water is conveyed through a trickle system of irrigation to vegetables planted in a controlled environment greenhouse of inflated plastic. Some difficulties have been encountered in holding temperatures down, but production of tomatoes and cucumbers has shown great profit potential.

Supported versus ground culture

Outdoors or in protected cultivation, tomatoes may be either supported (trellised or staked) or allowed to grow freely on the soil surface (called ground culture). Supports are commonly used in protected cultivation to make economical use of limited space. Moreover, spraying, harvesting, and certain other operations are easier when tomatoes are supported. This is especially important when hand sprayers are used. For intensive cultivation—for example, in home gardens and for production of off-season tomatoes where high fruit quality commands a premium price—the plants are always trellised. In the tropics, only in Colombia and the Philippines do growers use supports for extensive production of processing tomatoes.

In more advanced countries, tomatoes are usually grown with supports to obtain earlier, cleaner, and larger fruits, as well as to make field operations more convenient. In developing countries, supports are used principally to keep the fruits above the ground and to keep the plants from being blown down by strong winds, especially during the rainy season. But where pruning of stems is a standard practice, supported plants tend to lose more fruits due to sunburn, cracking, and blossom-end rot.

The materials and the labor to prop up the plants are enormously costly. The production of props (bamboo slats or wooden poles) has become a profitable minor industry in Mexico, Brazil, Colombia, Taiwan, and the Philippines. Trellising accounts for 13 percent of the total cost of production in Colombia, 30 percent in the Philippines, 25 percent in Taiwan, and 12 percent in Mexico.

Methods of supporting tomato plants vary from country to country (Figure 11). In Sinaloa, Mexico, for example, heavy poles are placed 4 to 5 meters apart and lighter poles are placed in between, every 50 to 100 centimeters. Ropes or wires are strung between these poles at different heights above the soil, and plants pruned to a single stem are tied to the poles or the

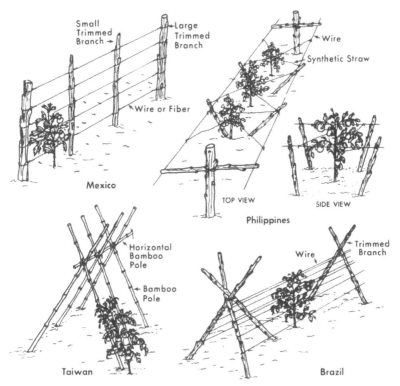

Figure 11. Systems used in the tropics for supporting tomato plants.

ropes. In Brazil, two heavy poles are placed in a slanting position on each side of the row so that the tops of the poles intersect and the point of intersection is tied securely. Pairs of poles are established every 3 to 4 meters, and four to five wires are strung at different heights from the ground. The plants are trained to the planting wires. The method used in Taiwan is similar, except that light bamboo poles are used instead of heavy wooden poles and an additional pole is placed horizontally in the V formed by the intersection of two or more poles. The poles are then tied for rigidity. In Claveria, the Philippines, a series of T poles are laid out every 5 to 8 meters. Wires are tied on each end of the horizontal bar of the poles. The plants are caught with strings, which in turn are tied to the wires. As the weight on the wires increases, a series of bamboo slats is placed in a V position and tied to the wires to provide additional support.

In greenhouses, a single stem is usually trained to a pole or rigid string. Pruning to a single stem by breaking off the lateral branches and tying the stem to a stake or pole is the most popular technique in U.S. greenhouses. Sometimes the plants are pruned to two or three stems. In other countries, however, most growers prefer a system of supporting tomato plants that is less costly than pruning.

It is safe to say that most of the world's outdoor tomatoes are raised in ground culture. The high yields possible with low production costs make ground culture attractive for extensive cultivation of processing tomatoes. If labor is expensive and large areas are to be planted, field operations can be fully mechanized, from drilling the seeds to harvesting. When growing conditions are very favorable, fresh-market tomatoes are generally raised in ground culture, too.

Transplanting and direct-seeding

Tomatoes are usually planted as seedlings (also called transplants). In affluent countries, however, some growers use such planting techniques as drilling seeds directly into the soil, direct-seeding with a plug mix, drilling a liquid containing pregerminated seeds, or laying tapes embedded with seeds over the field.

Growers of both fresh-market and processing tomatoes use transplants. Farmers and home growers usually raise their seedlings for 25 to 60 days. During this period the grower can carefully control conditions for growth and development so that more seedlings survive, which is especially important if expensive hybrid seeds are used. In addition, the seedlings can be held for a time, if necessary, before transplanting them.

In temperate countries, production of tomato and other seedlings is a lucrative nursery business. The seedlings are raised in hothouses during the spring for transplanting in the field after the danger of frost has passed. Operations in hothouses—such as seeding, fertilization, watering, and control of pests and diseases—are mainly automatic and require few laborers.

In Claveria, the Philippines, many growers of fresh-market tomatoes do direct-seeding by hand because labor is inexpensive. A large tomato-processing company in the same area also

Small tomato growers in the Philippines raise seedlings in makeshift sheds covered with palm fronds. (Source: PCARR.)

Automated spraying of seedlings in a greenhouse in Mexico. (Source: CIAPAN.)

uses this method. Laborers make three holes equidistant from one another (10 to 15 centimeters apart) and place three to five seeds in each hole. The holes are covered with rice hulls to keep the soil from caking and hardening. When the plants have three to four true leaves, they are thinned manually, leaving one plant per hole.

In countries where labor is expensive, large commercial growers use machines for direct-seeding, especially in production of tomatoes for processing, where high yields and low production costs are desirable. Direct-seeding requires three to four times more seeds than transplanting. In addition the farmer must be able to thoroughly prepare the land, control irrigation and drainage, and eliminate weeds. Seeds are drilled at a precise rate about 2 centimeters into the soil. When plants have two or three true leaves, they are mechanically thinned to one seedling per hill spaced about 15 to 25 centimeters apart or to clumps of two to four seedlings spaced 20 to 30 centimeters apart.

Some other methods of direct-seeding are becoming popular in the United States. The plug-mix method involves mixing tomato seeds, water, and fertilizer into a blended medium of vermiculite and peat. The mix is drilled in the soil at the rate of about 150 milliliters per hill. Fluid drilling, another method, involves germinating seeds and then mixing the best germinated seeds with a gel, which protects the seeds and suspends them for uniform distribution. The mix is sown through a special fluid-drill seeder. Because the seeds are pregerminated under ideal conditions and then only germinated seeds are selected for planting, the seedling stand attained in the field is more uniform than when seeds are sown dry.

Management practices

In tropical tomato production, practices for germinating seeds and growing productive plants are as important as good seeds. Practices used in temperate countries are rarely suitable, especially in the hot-dry or hot-wet tropics.

Growing seedlings. After a grower has obtained good seeds, the manner in which they are germinated and grown spells the initial success or failure of his tomato production. Low germination is often blamed on poor seed when the true cause is preemergence "damping-off." Seedlings are very susceptible to damping-off from the time the seed germinates until the seedling has at least one true leaf. Any environmental condition that retards the seedling development favors damping-off pathogens such as species of *Pythium, Rhizoctonia,* and *Fusarium.* Unfavorable conditions may be caused by soil mixtures that hold water too long or that have poor aeration, by the application of too much or too little water, and by improper fertilization. Many growers sow seeds thickly in boxes or beds that contain poorly drained soil and that are situated in a moist place with inadequate aeration. Taking steps to prevent damping-off is much easier than controlling it once it starts to develop.

A number of researchers advocate mixing rice hulls and soil to create a growing medium for tomato transplants. Rice hulls act as an inert material in the growing mixture. Horticulturists in the Philippines recommend mixing rice hulls and clay loam soil at a ratio of 2:1 by volume and adding, per liter of mixture, 1 gram of ordinary superphosphate applied before sowing, plus 3 grams of complete fertilizer (14-14-14) applied one-third before sowing, one-third about 10 days after sowing, and the rest about 20 days after sowing. This medium could be adopted in many areas where rice hulls go to waste.

Scientists at CIAPAN in Mexico have demonstrated that rice hulls could be substituted for an expensive imported growing medium that local growers have been using to produce transplants. Scientists at AVRDC have found that a satisfactory growing medium can be made by mixing 3 liters each of soil, compost, sand, and rice hulls and adding 1.8 grams of ammonium sulfate, 5.6 grams of calcium superphosphate, and 3.3 grams of potassium chloride. Fumigating or steaming the medium prevents soil-borne diseases.

Transplants can be raised in plant beds or in seedboxes filled with a medium such as a mixture of rice hulls and soil. There are many variations of these methods. In plant beds, seedlings are produced in mass and are generally less successful than seedlings grown in seed boxes, because diseases are hard to control and the seedlings are difficult to remove for transplanting without damage to the roots. These problems can be minimized in seed boxes by carefully spacing the seedlings. Seedlings grown singly are less likely to be damaged in transplanting because the medium will hold together during pulling and setting if the root system is fairly well developed.

Many farmers raise seedlings in other types of containers. Plastic containers that have individual cells filled with an appropriate medium are widely sold. Bands of paper glued together are satisfactory—the bands are stretched out on a wooden base and filled with a sterilized medium into which the seeds are sown. Rolled banana leaves can also serve as containers for individual seedlings instead of "jiffy" pots or small plastic pots, and they produce equally good transplants.

Regardless of the method used, young seedlings should be covered with nylon net to protect them from the intense heat of the sun, heavy rainfall, and strong winds. The net breaks the impact of falling raindrops and splashing soil particles, which can injure tender seedlings.

Nylon nets protect young seedlings from sun, rain, and wind. (Source: AVRDC.)

During the rainy season, high temperature and low light intensity may cause seedlings to be unhealthy and elongated, but these effects can be prevented by spraying 5000 ppm of Alar or 300 ppm of ethrel on the foliage when seedlings have two true leaves.

Transplanting. Tropical tomato growers usually do transplanting late in the afternoon, when the heat is less intense, to minimize transpiration of the seedlings, which permits them to recover faster. The following morning, the growers may cover the transplants with banana bracts, leaves of water hyacinth, or cardboard. The coverings are removed in the afternoon. For about a week the process is repeated daily to shield the transplants from the sun's heat.

Transplanting can be done at any time of day, however, provided seedlings are hardened by gradually exposing them to strong sunlight or by gradually withholding water from them. Hardening makes the plant tissue thicker, less succulent, and more resistant to transplanting damage than nonhardened plant tissue. The laborious task of covering and uncovering the plants is unnecessary if the seedlings have been hardened.

Inexperienced growers cause high seedling mortality by pulling the seedlings when the soil is dry, which destroys much of the roots. When these seedlings are transplanted, they must struggle to produce new roots at the same time they have undergone a drastic change in environment. High mortality also results from transplanting seedlings in a field that is too dry.

The best time to transplant properly grown and hardened seedlings is 21 to 30 days after they emerge. Ideally, water and fertilizers should be in the root zone of newly transplanted seedlings to improve seedling survival, enhance earlier recovery, and encourage faster growth. Each plant should receive about a quarter liter of water at transplanting. Furrow or sprinkler irrigation, if available, should start a day or two later. Water-soluble fertilizers (24 grams of 12–24–12 per 10 liters of water) are applied with the transplanting water. If water is plentiful the field can be irrigated four to six days before planting. In this case, a starter solution may be applied to the seedlings a week or two before transplanting.

The crew approach has been found to facilitate transplanting tomatoes. In a crew of four, one member digs holes, one distributes seedlings, one plants, and one irrigates. This system is very efficient in large operations.

Irrigation and drainage. Irrigation is highly desirable for growing tomatoes in the tropics. Even in the rainy season, rainfall distribution can be erratic—short periods with several centimeters of rain per day may be followed by periods with no precipitation. During the dry season, commercial tomato production is nearly impossible unless farmers have supplemental water available. If farmers have irrigation facilities, the proper rate and frequency of water application for tomatoes under a particular soil and climatic condition must be established.

Periods of heavy rainfall can be just as harmful as dry spells. In the rainy season growing conditions often are poor because excess water and poor drainage reduce soil aeration. Tomatoes, like most vegetable crops, require well-drained soil to maintain soil oxygen adequate for growth and development. Methods of controlling drainage and protecting plants from excessive rainfall deserve research attention.

In Taiwan and Thailand, many farmers overcome drainage problems during the rainy season by growing their vegetables in beds raised 30 to 50 centimeters. In Taiwan, beds are about 40 centimeters apart from edge to edge. In Thailand beds are about 1 to 1.5 meters apart because the ditch between beds serves as a passageway for small boats, which carry sprayers, other farm equipment, and farm produce. Above all, the ditch serves to carry off excess water. The beds vary from about 1 to 4 meters in width and from 5 to 50 meters in length. During the height of the rainy season, they look like small garden islands.

AVRDC scientists have found that in loamy and sandy soils, beds raised 30 centimeters, with compost (20 to 40 t/ha) in the center covered by rice straw mulch (10 to 20 t/ha), can reduce the effect of excess water during the rainy season and improve yields. The raised beds permit drainage; compost improves soil texture and aeration; mulching provides soil aeration and prevents compaction. Under rainy conditions, a sandy soil is more suitable for tomatoes than a loamy soil because it has better drainage and aeration, but fertilizer should be applied to compensate for the lower initial fertility of sandy soil and the easier leaching of nutrients.

Land preparation and tillage practices. Because tomato roots must have good aeration, proper soil structure is important. When rainfall is heavy, farmers have difficulty preparing land for planting. Working wet soil causes puddling (loss of soil structure) and formation of excessively large clods. Moreover, in the rainy season farmers have difficulty finding time to disk and harrow the clods to a proper size because frequent rains can make a field impassable. Following a heavy rain, land preparation in heavy soils often must be delayed several days, although many light soils can be worked within a day or two. The equipment used can make a difference, however. For example, plowing can be done with a buffalo, without destroying the soil structure, three to four days earlier than with a tractor.

Most tropical farmers use draft animals for tillage because they do not compact the soil and the clods formed by slow-moving animals produce better soil structure than those formed by heavy equipment. Animal-drawn implements can also prepare raised beds as well as tractors. But for preparing large areas, there is no substitute for tractors.

No-tillage techniques for growing tomatoes have been explored at the International Institute of Tropical Agriculture in Africa. A promising method involves establishing a cover crop of tropical kudzu (*Pueraria phaseoloides*), killing it with two applications of paraquat, and planting the tomatoes into holes dug through the layers of the cover crop, which serves as mulch. Digging holes is the only cultivation involved. The method has potential in areas where land preparation is difficult and mulching is essential for increased yields.

Fertilizer application. In the tropics tomatoes often respond poorly to fertilizer because farmers plant unadapted varieties and manage the soil improperly. If rainfall or irrigation is excessive, the pores in poorly prepared soil fill with water, restricting root development and lowering the response to fertilizer. These conditions discourage fertilizer use. Leaching of fertilizer in the rainy season is also a serious problem. Farmers need advice on when and how often to apply fertilizer under various soil conditions during both the dry and wet seasons.

Fertilizer trials with tomatoes during the rainy season often fail because unadapted varieties and inappropriate cultural practices are used. Recently, however, much progress has been made. At AVRDC's experimental farm, where the soil is a moderately well-drained silt loam, scientists have consistently obtained yields of 13 to 20 t/ha. They use improved varieties and apply fertilizer at a rate of 120–150–120 (N-P_2O_5-K_2O) with 20 t/ha of compost and 10 t/ha of straw mulch. Also, they have found that split application of the chemical fertilizer is better than one-time application. Before bedding, 40 kg/ha of nitrogen, all the P_2O_5 (150 kg/ha), and 60 kg/ha of K_2O are incorporated into the soil. Two weeks after transplanting, 40 kg/ha of nitrogen are applied and, four to five weeks after transplanting, the rest (40 kg/ha of nitrogen and 60 kg/ha of K_2O) is applied. Although these findings may be applicable only in moderately well-drained silt loam soil, similar studies could be conducted in other areas or soil types to generate information on tomato fertilization during the wet season.

Mulching and other cultural practices. Mulching with rice straw almost always raises yields because it prevents soil erosion, especially on lighter soils. It also minimizes soil compaction, slows weed growth, and prevents contact between the fruit and damp soil. In the lowlands, determinate tomatoes can be grown at lower cost and with better results on mulched plots than with trellising.

Mulching with plastic is inappropriate for the lowlands because it raises soil temperature and interferes with soil aeration. In the highlands, however, plastic mulching gives high yields (Figure 12), primarily because fewer fruits are lost, though the increase in soil temperature also favors the growth and development of the plants.

The use of black plastic mulch costs much less in labor and materials than the use of wire trellises, but if indeterminate tomatoes are grown they should be trellised. Without trellises their sprawling growth hampers control of diseases and insects, especially during the rainy season, and makes harvesting difficult.

Weed control. Weeds lower tomato yields by competing for light, water, carbon dioxide, and soil nutrients and by serving as alternative hosts for insects and diseases. The damage done

In the lowlands, growing tomatoes in raised beds covered with straw mulch can improve yields during the rainy season. (Source: AVRDC.)

In the highlands, plastic mulch can substitute for straw mulch. (Source: Twin Rivers Research Center, Philippines.)

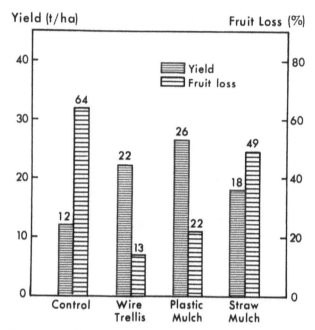

Figure 12. Effect of trellising and mulching on the yield of tomatoes under highland conditions. (Source: Twin Rivers Research Center, Philippines.)

by weeds is greatest during the wet season, probably because they grow faster than tomatoes when there is excess water. Many progressive growers in the tropics control weeds with chemicals, but in home gardens and small tomato fields, hand pulling, hoeing, and mulching are best.

Cultivation and the use of chemicals can be recommended in large operations where labor is scarce and expensive. Chemicals are also useful when wet fields make hand weeding difficult. AVRDC scientists have found that applying Lasso with a hand sprayer (2 kg/ha, active ingredient) at the time of bed preparation, when the soil is relatively wet, can control weeds for three weeks after planting. A follow-up spraying with paraquat (0.5 kg/ha) three and six weeks after planting can minimize weed problems during the rainy season. Since paraquat is a contact herbicide, a nozzle shield must be used to protect the tomato seedlings from severe damage.

It is important to control weeds at the right time. Weeds cause the most damage during the first 25 to 30 percent of the crop's growth. Thus, in tomatoes, weeding should be done from transplanting to early fruiting stage.

Disease control. Although severe attacks of some diseases can destroy a tomato crop, use of resistant varieties, good agronomic practices, and appropriate chemicals will minimize the potential damage. If soil-borne problems such as bacterial wilt, fusarium wilt, and root-knot nematodes are common, resistant varieties must be planted because chemical control is expensive and not wholly effective. Cultural practices such as the planting of marigolds and the application of chicken manure can help to suppress root-knot nematodes.

Many leaf diseases can be kept under control with appropriate chemicals applied frequently and at the right time. Benlate (0.14 kg/ha) and Manzate D (0.6 kg/ha), sprayed alternately, will control the most common leaf diseases. About 10 sprayings per cropping season are necessary.

Insect control. Crop rotation, insect repellants, and disposal of crop residues in the fields are cultural practices that can minimize insect damage in the tropics. Systemic insecticides, which are absorbed by the plant and translocated from treated to untreated parts of the plant, are more effective than conventional insecticides, which are likely to be washed off the leaves by heavy rain. At AVRDC, spraying Lannate (0.14 kg/ha) 8 to 10 times per cropping season controls most insects that feed on tomatoes.

Some relatives of the tomato that are less severely attacked by insects have been identified by U.S. scientists. *Lycopersicon hirsutum,* for example, has resistance to leaf miner, tobacco flea beetle, greenhouse whitefly, Colorado potato beetle, and two species of spider mites. Nevertheless, there have been no concerted efforts to develop insect-resistant tomatoes.

An outstanding example of integrated pest management is the control of vegetable leaf miner, *Liriomyza sativae,* in tomato crops in one area of Florida. The basis of the control is maintaining populations of wasps that are parasitic to leaf miner by minimizing application of insecticides. To do so, the population levels of the leaf miner are surveyed twice a week. Only if the pest levels become high are chemicals applied, and then only selective insecticides or broad-spectrum insecticides at low rates are used. By eliminating unnecessary insecticide applications, this system reduces production costs.

Postharvest Technology
and Marketing

For vegetables, postharvest handling is as critical as production practices. In fact, postharvest waste is often greater than
production losses in developing countries. A study in the Philippines found that losses due to insect pests, diseases, and weeds
amount to a quarter of total vegetable production, but postharvest waste (decay, overripening, mechanical injury, weight
loss, etc.) was even higher—over 40 percent.

Tomatoes are especially vulnerable to postharvest waste.
Common estimates of postharvest loss in the tropics range from
5 to 50 percent. To minimize losses, tomatoes must be harvested
at the right time, because overripe tomatoes are more susceptible to physical injury than ripe and pink ones. They must be
packed properly to avoid injury that would cause softening and
decay.

Harvesting

For processing, tomatoes are harvested red ripe and for the
fresh market they are harvested at various color stages. The
availability and cost of labor, distance to market, intended use,
and factory capacities all influence the manner and number of
harvests.

Labor. If labor is inexpensive, processing tomatoes are
harvested 4 to 11 times per cropping season. Fewer harvests are
made if labor is scarce and expensive. In developed nations, a
single harvest that destroys the plants is increasingly common.
For machine-harvesting, the tomato variety must produce a

concentration of fruits that mature at the same time so that in one harvest the highest quantity of marketable tomatoes can be obtained.

For fresh-market tomatoes, growers ordinarily make 4 to 15 harvests. (Although in Western European greenhouses, where tomato plants grow for about 12 months, up to 30 harvests are made.)

Distance to market. When production areas are close to market, farmers harvest frequently because it is easy to continually supply markets with fresh tomatoes. In developing countries where markets are distant, most fresh-market tomatoes are harvested at the mature green stage or the breaker stage to reduce losses in quantity and quality caused by poor transportation and careless handling. Nevertheless, in some countries—particularly in Africa and Latin America—farmers harvest very ripe tomatoes, which have special uses in local food dishes.

Color. For processing tomatoes frequency of harvests (either by hand or machine) is largely determined by how fast the fruits turn red ripe. For machine harvesting, however, a concentration of red ripe tomatoes is important because only one harvest is made.

Fresh-market tomatoes are harvested red ripe, breaker, or mature green, depending on the desired color in the market. In North America, however, tomatoes should appear red ripe when displayed in markets, so tomatoes harvested mature green or breaker are treated with ethylene to induce even ripening.

Packing

Proper packing protects tomatoes from bruises and blemishes while in transit to consumers or factories. A cut or broken tomato may easily harbor a pathogen that may spread and spoil all tomatoes in a lot. If a factory checks mold counts, one spoiled tomato could result in the rejection of a whole shipment. If rotten parts must be removed, labor costs increase.

Proper packing of processing tomatoes is relatively simple. They are put in wooden or plastic crates or bamboo baskets after harvest and brought from farms to the factories. In the United States, machine-harvested tomatoes are put in large bulk containers, which may be fiberglass gondolas (holding several thousand kilograms) or plywood bins (holding several hundred kilograms); loaded on huge trailers; and delivered to the factories.

In the tropics, fresh-market tomatoes are commonly packed in wooden crates or in bamboo baskets that accommodate 10 to 30 kilograms. Rigid plastic or cardboard boxes are used in many developed countries. Newly harvested tomatoes are usually collected in a packing shed or at a nearby road where transactions are made.

Under the best postharvest handling practices, tomatoes are first washed, waxed, and then graded before being carefully packed in containers. The washing water, however, can be a source of contamination. It should be chlorinated to prevent bacterial infection and heated ($3°$ to $6°$ C. higher than the temperature of the tomatoes) to prevent the tomatoes from absorbing it. In some instances, the tomatoes are transported in the containers originally packed in the field or they may be repacked in other containers before transport.

Hauling

In developing countries, tomatoes are carried in baskets, in bullock-drawn sleds, on the backs of horses or donkeys, in trucks and jeeps, and in many other ways. When production areas are near rivers, they are carried in dugout canoes and motor boats. In most developed countries, the hauling is done by trucks and trailers pulled by tractors.

Processing tomatoes are generally taken directly to factories in various types of containers where they await grading prior to processing. Especially in developing countries, tomatoes are sometimes hauled to a shed where severely cracked, worm-damaged, or poor color fruits are culled before delivery to the factories.

Storage

In Taiwan, processing tomatoes are stored in the open at the factory site. Sometimes, however, the factories force farmers to practice "vine storage" by limiting the number of crates or boxes they will accept each day. Vine storage means the earliest maturing fruits are left on the vine for several weeks after they reach full color. The tomatoes have to be firm-fruited to remain in good condition on the vine after they reach full color. In the United States, vine storage of tomatoes is common in California, especially during the peak processing season.

In developing countries, fresh-market tomatoes are usually stored overnight in the farmer's home or packing shed. In some developed countries, to initiate ripening, they are stored in temperature-controlled rooms where they are treated with ethylene gas for 24 to 72 hours before being shipped. Ideally, green tomatoes should be transported at 13° to 18° C. and 85 to 90 percent relative humidity. Recommended transit conditions for pink tomatoes are about 7.5° C. and 95 percent relative humidity. After the tomatoes arrive at their destination, they are usually repacked in small packages and then placed in refrigerated display cases. The optimum range of temperature for tomato storage is 18° to 22° C. Great care in storing fresh-market tomatoes is necessary to maintain their quality.

Practical methods of postharvest handling

Postharvest losses in the tropics due to physical, physiological, and pathological damage to tomatoes can be reduced in several ways.

Appropriate varieties. Varieties differ in their susceptibility to bruising and decay. They also vary in their rate of ripening. Although no variety yet combines all desirable postharvest characteristics, the postharvest quality of varieties that are available should be compared to guide selection of varieties to be grown.

Maturity at harvest. Degree of ripeness at the time of harvest is an important aspect of the keeping quality of tomatoes. Mature green tomatoes, for example, are less vulnerable to bruising and breaking of the skin than pink, ripe, or overripe tomatoes, in that order. Mature green tomatoes also possess longer shelf life, although their eating quality is usually considered inferior to that of riper fruit.

Gentle handling. Because of the soft texture of tomatoes, they should be handled gently to avoid bruising and breaking the skin. Slight bruising creates favorable conditions for decay organisms and stimulates deterioration and dehydration of the fruits. Since the pedicel (stalk) of a fruit can easily puncture the skin of neighboring fruits, removing the pedicels lowers postharvest losses. Soft packaging materials (paper, banana leaves, etc.) and rigid containers also reduce damage.

Temperature control. Cooling extends the storage life of horticultural products by slowing physiological changes. Refrigeration is the most common method, but even simple methods such as harvesting during the cool early morning or late afternoon hours will cut losses caused by high temperatures. Similarly, keeping tomatoes under shade prolongs storage life. Exposure to direct rays of the sun is comparable to cooking the tomatoes because their internal temperature rises much higher than the air temperature.

Shortening time between harvest and consumption. Extending the harvest season by growing varieties that mature at different times, by staggering planting dates, or by improving transportation and marketing facilities leads to more orderly marketing and shortens the time between harvest and consumption. The first two methods are simple and do not involve great expense, but the third usually requires government action and large sums of money.

Domestic marketing

There are no typical marketing channels. They may range from simply selling the tomatoes to passersby along a road to bringing them to an auction center where they pass from one handler to another before reaching the ultimate consumer.

Developing countries. The Sundarapuram Vegetable Assembly Center in Coimbatore, India, is an example of one kind of marketing channel (Figure 13). It is owned and operated by the local government and farmers as much as a hundred kilometers away bring their produce by trucks and bullock-drawn carts. Since the auction starts at 5 a.m., farmers deliver tomatoes from midnight until early morning.

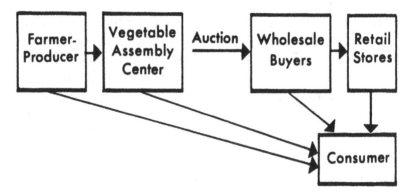

Figure 13. Flow of fresh-market tomatoes through Sundarapuram Vegetable Assembly Center in Coimbatore, India. (Source: Tamil Nadu Agricultural University.)

Tomatoes are auctioned in baskets of 10, 15, and 30 kilograms. The auctioneer serves as the middleman for the farmers and receives about 8 percent of each sale. Farmers have to pay the assembly market about $0.01 per basket of tomatoes regardless of size. Payment for the auctioned tomatoes is given to the auctioneer who, in turn, collects his share and gives the balance to the farmers. The volume of tomatoes traded in this center ranges from 600 baskets per day during the off-season to 5000 baskets per day at the peak of the tomato-production season.

Few of the tomatoes purchased at the center go directly to local consumers. Most buyers repack newly purchased tomatoes in wooden crates for shipment to distant destinations. Before repacking, they are sorted according to size and degree of spoilage. Large and good quality tomatoes are sent to the big cities for distribution to retail stores; the rest go to nearby retail markets.

In Bangalore, India, both fresh-market and processing tomatoes are marketed. The flow of fresh-market tomatoes is similar to the one in Coimbatore (Figure 14). The marketing of processing tomatoes is relatively simple. They are delivered to a factory that has contracted with certain farmers and payments are made at the factory on delivery. If the variety grown is one recommended by the state or central government, private or government seed companies may contract with farmers for some of the seeds from the fruits. To ensure the purity of the seed lots, government seed inspectors visit tomato fields and

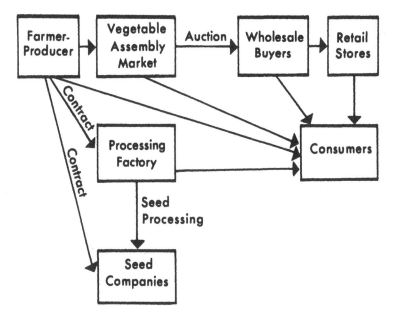

Figure 14. Flow of fresh-market, processing tomatoes, and tomato seeds in Bangalore, India. (Source: Indo American Hybrid Seed Company.)

rogue (pull out) off-types and diseased plants. At the factory, a special sieve is used to separate seeds from the pulp. Seeds with tomato juice are fermented overnight and washed and dried the following day. Field inspection and seed processing are arranged by the seed contracting agency. A tomato farmer in Bangalore, therefore, gets income from selling both fruits and seeds. This arrangement also enables seed distributing agencies to maintain a supply of recommended tomato varieties.

In Taiwan, farmers generally harvest tomatoes in the afternoon and most do some sorting and grading before selling the next day. Graded tomatoes are then packed in cardboard cartons (30-kilogram capacity) with the name and address of the producer and net weight and grade of the produce. However, some tomatoes are still sold ungraded in bamboo crates to buyers who may do the grading later.

The major outlets for the fresh-market tomatoes are local shippers, wholesale markets, and retailers. The bulk of the tomato crop in central Taiwan is sold by farmers to wholesalers and shippers. They forward the tomatoes to the terminal market in Taipei, with the remainder going to other urban areas. Producers on the fringes of large urban areas or in concentrated vegetable growing locations in rural areas sell either to local retailers or to a shipper who has a local collection point.

In minor production areas, it is often not worthwhile for producers to transport their tomatoes to retailers or wholesale markets. Instead they prefer to sell at a modest price to merchants who collect tomatoes and transport them to large wholesale markets. In major vegetable production areas, however, producers usually transport their tomatoes to large wholesale markets.

Wholesale markets are found in production areas as well as in population centers. Producers and shippers sell their tomatoes in wholesale markets, but wholesalers, exporters, and shippers buy and transport tomatoes to urban markets, institutions, and foreign countries. The volume of sales in these markets is seasonal.

In a typical assembly market, the buyer registers and deposits money with the cashier to get a license that entitles him to buy vegetables. Only licensed buyers can transact business at a market. A farmer delivers his tomatoes to the market where they are weighed and appropriate labels attached to the baskets. He gets a receipt if he wishes to leave the produce at the market and collect his money the following day. If he decides to stay, he waits for a buyer. Once the sale is completed, the farmer collects payment from the cashier by presenting the license number of the buyer.

Developed countries. In packing plants in developed countries, tomatoes pass through assembly lines that include mechanized units for washing, sorting, grading, and packing. The tomatoes are graded according to size, color, and amount of blemish on the fruits. Grading information appears on the boxes, especially if the tomatoes are for export. Tomatoes are packed in 14-kilogram boxes and later repacked in consumer packs or prepacks, which come in various sizes. Prepacks are sometimes prepared by the supermarket chains. Prepacks are covered with plastic film that is permeable to gases but not to water.

Tomatoes are usually conditioned in cold storage before they are placed in refrigerated trucks for shipment to long-distance markets. Upon arrival they are kept in cold storage until ready for repacking and display in the supermarkets. For nearby markets, however, tomatoes are often shipped in trucks that are not refrigerated.

During the harvesting season, tomatoes are commonly sold locally to customers at roadside stands. Also, in parts of the United States, some consumers go to farms and harvest tomatoes themselves at a low price.

Export marketing

Most importing countries establish quality and grade standards for both fresh-market and processed tomatoes. For this reason, before going into the export business, the first step is to study the importing countries' standards and special requirements, such as mold counts for processing tomatoes, pesticide tolerance levels, and pesticides prohibited in the production of tomatoes.

The basic requirements vary considerably among importing countries. For example, Singapore receives ungraded tomatoes from Malaysia, Indonesia, Thailand, and Taiwan. Tomatoes arrive in bamboo crates and cardboard or wooden boxes, sometimes without any information on variety, producer, or origin. Tomatoes exported to Japan or the United States, however, must meet strict packaging and labeling standards.

Mexican growers have been using U.S.–bred tomatoes and follow U.S. and Canadian grade standards, labeling requirements, and pesticide regulations. Similarly, Moroccan growers adhere to European standards for their tomato exports, and Taiwan meets the various grade requirements for tomato products for export to Japan, Canada, the United States, Saudi Arabia, and other countries.

8
Promising Future Research

Research can diminish the risks of growing tomatoes in the tropics. Breakthroughs in certain areas will make the crop more productive for farmers and more available in tropical markets. The scope of research that developing countries can usefully undertake is vast. For efficiency, more fundamental studies are better handled by international centers and universities in affluent nations.

At the Asian Vegetable Research and Development Center tropical tomatoes are a major area of research. Several other international agricultural research institutes (see Appendix A) conduct cropping systems research, which embraces a wide variety of vegetables. These centers, as well as scores of universities and experiment stations in developed countries, are valuable sources of germ plasm and information on vegetable crops. National vegetable programs in some tropical countries are associated with these centers and universities; other programs could benefit by making similar arrangements.

Developed countries such as the Netherlands, Belgium, and Norway have found there is a big payoff from research on a high-value crop like tomatoes. Government institutions as well as private seed companies in those countries are engaged in developing varieties and technology for greenhouse tomato production. Their efforts have been rewarded by high yields per hectare, plentiful supplies for domestic use, and sufficient quantities for export to neighboring countries. Countries that have at least a thousand hectares devoted to tomatoes can justify allocating funds to form a core staff for research on this crop.

The topics discussed below should get high priority from tomato researchers in tropical countries.

Varietal improvement

High yield potential is a goal of all tomato-breeding programs, but it deserves added emphasis in developing countries where little or no premium is paid for quality. As consumer incomes grow, quality will warrant more attention. Yield potential can be raised by breeding varieties for disease resistance and good fruit-setting ability under hot-dry and hot-wet tropical conditions.

Resistance to diseases. A number of bacterial, viral, and fungal diseases and root-knot nematodes can seriously affect tomatoes. Before doing any breeding, researchers should categorize the nation's tomato diseases according to their importance. To be successful in the tropics a variety usually needs resistance to bacterial wilt, two or three virus and fungal diseases, and root-knot nematodes.

The most serious disease of tomatoes in the tropics is bacterial wilt. Varieties or breeding lines that are resistant in one country may be susceptible elsewhere. Some AVRDC breeding lines hold resistance at several locations, but the degree of resistance is influenced by the plant's genotype, the pathogen, and the environment. In fact, from a given cross, desirable selections may differ from country to country (Table 11), primarily because of differences in strains of bacterial wilt, temperature fluctuations, and soil conditions. Nevertheless, it is possible to obtain a genotype with wide adaptability. (In Table 11, "SSD 12," which gave the highest yield both in Taiwan and Malaysia, is an example.)

National tomato programs can benefit from evaluating local or imported varieties and breeding lines in the general area where the tomatoes will be eventually grown commercially. In evaluating wilt-resistant materials, it is imperative to spray them with chemicals to control leaf diseases as they may be resistant to wilt but susceptible to a variety of leaf diseases. If breeding is to be done locally, progress will be faster if resistant advanced breeding materials developed elsewhere are used

Table 11
Differences in the strains of bacterial wilt, temperature fluctuations, and soils in three countries cause differences in yield response of tomato lines, even though they have the same parents (VC8-1-2-1/Venus//Kewalo).

Taiwan		Malaysia		Papua New Guinea	
SSD no.	Yield (t/ha)	SSD no.	Yield (t/ha)	SSD no.	Yield (t/ha)
12	20	12	36	22	48
27	20	47	24	15	42
15	20	16	24	18	38
14	19	18	24	24	34
Green Fruit (check)	1	Red Cloud (check)	0	(no check)	

Source: AVRDC, 1979. *Proceedings of the 1st International Symposium on Tropical Tomato,* Shanhua, Taiwan.

instead of starting with wild species of *Lycopersicon.*

Because resistance to bacterial wilt is complex (horizontal resistance), it takes a long time to develop a resistant variety. If a simply inherited type of resistance (vertical resistance) could be found, it would facilitate incorporation in adapted varieties. There is evidence that both types of resistance exist. For a short-range national program, therefore, vertical resistance is a sensible goal. For the more stable resistance needed in a long-range program, however, plant breeders should aim to incorporate horizontal resistance to bacterial wilt. A combination of both types of resistance would surely be of lasting benefit to farmers.

Unfortunately, resistance to bacterial wilt appears linked to small fruit size. At AVRDC, scientists grew 4000 F_2 genotypes from a resistant-susceptible cross and obtained not one resistant selection that had large fruits. In areas where large fruits are highly desirable, farmers may be slow to accept wilt-resistant selections. This problem needs attention.

Various virus diseases attack tomatoes, often simultaneously. The most common is TMV (tobacco mosaic virus). Some sources of resistance are known, but resistance seems associated with defects such as small fruits, poor fruit-set, horizontal growth, hairless stems, and a tendency for fruits to be excessively succulent. Nevertheless, TMV-resistant varieties have been adopted in many greenhouse production areas.

In outdoor production the advantage of using TMV-resistant lines is mostly nullified by the presence of other virus diseases. For example, in the presence of potato virus X, TMV-resistant plants succumb to a disease called double-streak virus. When tobacco etch virus, potato virus Y, and cucumber mosaic virus are present, TMV resistance usually makes little difference. When whitefly-transmitted leaf curl virus and thrips-transmitted spotted wilt virus are present, TMV-resistant plants are useless. Combining resistance to the two or three most harmful local virus diseases should be a major research goal.

Unlike bacterial and virus diseases, most fungal diseases of the leaves can be controlled by chemicals. But chemicals are generally uneconomical in the tropics, so the only practical solution is a long-range program of breeding for resistance to the most serious fungal diseases, such as early blight, gray leaf spot, leaf molds, and septoria leaf spot. For highland production, late blight resistance and resistance to two soil-borne pathogens—fusarium and verticillium wilt—should also be incorporated. Sources of resistance to these diseases have been found, but the ease of transferring the resistance genes varies from one disease to another. Breeding for late blight resistance is challenging because pathogenic races and environmental conditions greatly influence the disease reaction of the plants.

Root-knot nematodes tend to flourish in soils where tomatoes are grown. The single dominant gene *Mi*, which governs resistance to root-knot nematodes, is useful to incorporate into susceptible local varieties. Even though it is associated with some undesirable traits and has given variable results in regional testing, it has saved tomato growers millions of dollars in some areas. Researchers in national programs should survey a wide range of tomato accessions for nematode resistance because at present all nematode-resistant varieties have the same *L. peruvianum* parent as the source of resistance.

Tolerance to heat and moisture. Disease resistance is worthless if the tomato plants fail to produce fruits. Except in the highlands, tropical areas rarely have the cool night temperatures desirable for fruit-setting.

Varieties that possess the ability to set fruits at high temperatures also need moisture tolerance, because in much of the tropics the period of high temperature coincides with the period of high rainfall. Some breeding materials from AVRDC possess these traits, but they should be more widely evaluated in tropical countries. The identification of 39 heat-tolerant accessions from 15 countries suggests that diverse heat-tolerant genes exist. National programs can take advantage of these materials and countries with sufficient funds and qualified plant breeders could also transfer genes for heat and moisture tolerance to already adapted local varieties.

The inheritance of heat tolerance is complex. It would be advantageous if a simply inherited gene for this trait could be identified. Such a gene, if dominant, could easily be incorporated and it would permit production of hybrid heat-tolerant materials.

Scientists at AVRDC have observed that fruit-setting score (1 to 5 in increasing order of fruit-setting intensity) is highly correlated with fruit-setting rate (ratio of number of fruits to number of flowers), which suggests that high fruit-setting score could be used as an indicator of high fruit-setting ability. Thus, if tomato lines can be planted in the field during the summer when rainfall is heavy, fruit-setting score can be used to screen for heat tolerance. Before national programs adopt this technique, researchers should verify it under local conditions. Since growing tomatoes to maturity in the field and then scoring them is laborious and time consuming, it would be worthwhile for researchers to attempt to devise other screening techniques for heat tolerance. A reliable way to screen in the seedling stage would be especially useful.

Good quality. Because of the numerous colors, flavors, sizes, and shapes of tomatoes that are possible, no single variety can satisfy all preferences. A researcher can set broad quality goals for breeding fresh-market tomatoes by investigating consumer preferences for acidic versus sweet tomatoes, pink versus red tomatoes, and the like, in the region where they will be consumed. In Sri Lanka, for example, more acidic tomatoes are preferred for curried dishes, so they are called curry tomatoes. The sweeter ones are called salad tomatoes.

For processing tomatoes, the breeder must aim to combine such qualities as desirable pH, soluble solids, viscosity, and good color with high yield potential.

The value of attempting to improve beta-carotene and ascorbic acid contents—the two important nutrients in tomatoes—depends on the priority national planners give to nutrition. Both traits are simply inherited. However, the color of high beta-carotene tomatoes tends to be more orange than many consumers like.

The three ripening mutants mentioned in Chapter 5 should be explored to improve the storage life of tropical tomatoes, thereby improving the quality of tomatoes in the marketplace.

Early maturity. Early maturing lines with high yield potential and strong resistance to pests and diseases are important in cropping systems that involve growing several crops each year on the same piece of land. Such lines can also reduce the farmer's risk in the rainy season because they are exposed to damage from high winds and excessive rainfall for a shorter time. Similarly, if water supply is limited, early maturing lines are less likely to be exposed to drought. They also need less fertilizer, pesticides, and labor because they occupy the land for a shorter time.

Freedom from fruit-cracking. Heavy rainfall increases the incidence of cracks, which make the fruit unsightly and serve as points of entry for fruit flies and rotting pathogens. Tropical tomatoes need some resistance to fruit-cracking to reduce spoilage and fruit loss. Local plant breeders can take advantage of breeding lines released by temperate countries if the desired traits cannot be found in domestic varieties.

Management practices

Fertilizer. The amount and kind of fertilizer to be applied depend upon the requirements of the tomato variety and the amount of nutrients the soil can supply. Soil type (sandy loam, clay loam, silt loam, etc.) largely dictates the soil's capability for supplying and retaining nutrients. A fertilizer recommendation for a specific variety and location should be formulated by researchers in the area where the tomatoes will be grown. Field experimentation, soil testing, plant analysis, and expression of deficiency symptoms are used singly or in combination to determine appropriate fertilizer recommendations.

As more locally bred varieties are produced, national soil scientists should study not only the amount and kind of fertilizers for high yield but also the right balance for high-quality tomatoes. Scientists in the temperate countries have found, for example, that nitrogen sources are of minor importance as long as input levels and yields are low. For high yield and good fruit quality, however, nitrogen sources are very important. Potassium affects not only tomato yield but also fruit quality and disease resistance. Its interaction with other nutrients should be studied in tropical soils. Attention should be paid to the most efficient methods of fertilizer placement and timing in locations where tomatoes will be grown, as well as foliar application of some nutrients. The importance of organic matter (farm manure, compost, green manure, etc.) in relation to water and nutrient availability should be closely studied, too.

Protected cultivation. In the lowland tropics, growing tomatoes during the rainy season can be made less risky by protecting the plants with plastic houses or high- or low-level tunnels. But the varieties grown must be disease-resistant and heat tolerant and the structures must be appropriately designed.

Research should be conducted on design, roofing materials (plastic film, fiberglass, and glass), and structural materials. Fiberglass and glass are expensive but durable. A plastic film that is not easily degraded by ultraviolet rays from the sun and that will last for at least two years makes a good roofing material. The houses should be built in relatively storm-free areas to minimize the chances of destruction by strong winds. Because they can be built with open sides and with means for warm air to escape through the roof, they do not need a cooling system, which would require expensive energy.

Scientists must find suitable management practices for tomatoes grown under protection in hot climates. The problems of frequency and timing of applications of fertilizers, pesticides, and irrigation water differ from those in the temperate countries and even from those in the highlands of the tropics.

Insect and disease control. The most economical way to control pests and diseases is through the development of resistant varieties, but integrated pest management—combining the use of resistant varieties with the application of chemicals, good cultural practices, and steps that favor natural enemies of harmful insects—is more effective.

Scientists in national programs should direct some research attention to the timing and frequency of chemical application, screening for the right chemicals, use of stickers to keep chemicals from washing off the plants, and use of systemic chemicals, if available. Studies of the seasonal distribution of pathogenic spores and insect populations would make planting and spraying schedules more effective.

Scientists in the Philippines have raised the possibility of using cross protection and antagonistic microorganisms to control bacterial wilt. Cross protection is an immunological response: a plant that shows an immune reaction to a virulent strain of a disease after being exposed to a mild strain is said to exhibit cross protection. The idea of using antagonistic microorganisms against bacterial wilt is based on the fact that certain microorganisms that live in the root zone of the tomato plant are hostile to wilt bacteria. Practical ways for farmers to inoculate seedlings with a mild strain of bacterial wilt before transplanting or to increase the population of soil microorganisms that are antagonistic to wilt pathogens would be valuable weapons against bacterial wilt, in addition to the development of resistant varieties.

More research is also needed on the use of natural enemies of insects and on cultural practices to reduce losses due to pests and diseases. In the Philippines, for example, scientists found that intercropping cabbage and tomatoes reduced the population and egglaying activity of the most destructive insect pest of cabbage, the diamondback moth (*Plutella xylostella*). The phenomenon was attributed to volatile compounds emitted by tomato plants. Studies of other crop associations might reveal relationships in which tomatoes would benefit. Crop rotations involving tomatoes are another subject that requires research. In Senegal, for example, scientists recommend following tomatoes with peanuts to control root-knot nematodes.

Interest in using insect pheromones (hormones) or similar

compounds against pests has been increasing in recent years. Pheromones have been demonstrated for more than 100 insect species. Some scientists have found that simultaneous evaporation of certain chemicals can disrupt the premating communication of several species attacking crops in an area. Pheromones are a potentially important component of integrated pest control.

Weed control. For growers with a few hundred square meters of tomatoes, the best weed control is hand pulling and cultivation with plows and hoes. For larger areas, a combination of physical and chemical means becomes a necessity, particularly when the fields are too wet to be tilled by mechanical implements. Knowledge of the weed species, their seasonal distribution and life cycle, can contribute to effective control. Finding the right herbicide and an application rate that will kill or stunt weeds without harming the tomato plant should be another important research undertaking.

Land preparation and tillage practices. Researchers must develop tillage and management practices for optimum tomato production during both the wet and dry seasons in the tropics. How high to construct beds to provide drainage in the rainy season and the best ways to furrow irrigate leveled fields in the dry season are examples of critical research topics.

Problem soils

The tropics has millions of hectares of problem soils—soils with low levels of phosphorus, zinc, and other elements; highly acid soils; and soils with toxic levels of elements such as iron and aluminum. Soil scientists can evaluate soil amendments that will reduce the harmful effects of these problems and plant breeders can screen for tolerant varieties. Better understanding of these problems would help tomato growers, as well as other farmers.

Intensive or multiple cropping

Farmers in many tropical areas practice intensive cropping, or multiple cropping. In Taiwan, farmers intercrop tomatoes with sugarcane, mangoes, and papaya.

Many small sugarcane growers could intercrop the spaces between rows of sugarcane plants and increase their income. It would also be worthwhile to evaluate the potential of intercropping tomatoes with fruit trees and other permanent crops. For example, millions of hectares are devoted to coconut farming. Spaces between coconut trees usually have sufficient light to grow tomatoes, but farmers need sound technical advice on suitable practices.

Scientists at AVRDC have examined the possibility of developing tomato varieties for relay cropping (planting before a preceding crop is harvested). They have found wide variation in the ability of tomato varieties and breeding lines to perform as a relay crop. In general, relayed tomatoes yield substantially less than those grown in monoculture, but some types perform equally well in both types of culture, suggesting it is possible to develop tomato varieties for relay cropping. Scientists must identify or breed special varieties for this purpose and determine the most suitable management practices.

Postharvest technology

Inefficient transportation and poor roads from farms to markets make postharvest losses of tomatoes high. The losses are accentuated by improper packaging and handling. Methods might be found to improve existing containers through the use of local materials, such as bamboo and lumber. The use of simple and inexpensive packaging materials should be explored as a means to lengthen the shelf-life of tomatoes, which are normally displayed in open stalls or kept refrigerated at home. For exports, packaging and containers must be designed for long trips by trucks, boats, or airplanes. Also, local grades and standards should be established so that a premium can be paid for quality. (Export tomatoes must conform to the grades and standards of importing countries.)

Seed production and distribution

In most tropical countries, agricultural universities and

government experiment stations are responsible for producing and selling seeds to farmers. Experiences in temperate countries, however, show that private seed companies do a better job of seed production and distribution. In tropical countries, the government can conduct seed production and technology studies that may encourage private enterprises to enter the business. It may be helpful if the government provides private enterprises with tax incentives and access to breeding materials and information. The government could gradually shift to basic genetic studies and seed production of crops that private companies do not handle. Tomatoes are an attractive crop for a private seed company, especially if hybrid seeds become locally acceptable.

Mechanization

Although it may be unnecessary to mechanize most cultural operations in tomato production in the tropics because low-cost labor is available, conditions may change as more factories are built and employment increases. In Taiwan, for instance, labor is getting more expensive and mechanical harvesting of processing tomatoes may be the only way for Taiwan to stay competitive. Research on design of small tomato harvesters should be undertaken because most farms in the tropics are small. Other small farm equipment should be developed for rainy season planting and other farm operations. National research programs may not be in a position to carry out such studies, but they could cooperate with appropriate institutions on this matter.

Causes of poor fruit-set

Yield per hectare, the ultimate measure of tomato productivity, is influenced largely by the number of good quality fruits that set. Fruit-setting is a complex physiological phenomenon that probably starts as soon as the seed is sown. When conditions (water, nutrients, and temperature) are appropriate, the tomato seedling grows vigorously, produces flowers, and sets fruits. Most, if not all, of the processes going on in the plant are temperature dependent. There is a narrow range of optimum temperatures for germination, seedling growth, flower initiation, flowering, fruit-setting, and fruit ripening. High temperature deters successful tomato production in the tropics. Moreover, excessive moisture in the air and in the soil greatly affects production.

Many intriguing questions about fruit-setting are unanswered. How long should tomato flowers be exposed to cool temperatures to have optimum fruit-set? Would a short dash of cool temperature be sufficient to affect fruit-setting in a manner similar to a short dash of light that brings a long-day plant to flower? Would flower thinning increase fruit-setting?

Efficient research organization

An interdisciplinary approach to research pays dividends because most problems are complex and they nearly always exceed the boundaries of any one discipline. Individuals on interdisciplinary scientific teams should be chosen for their ability to cooperate.

The Philippines provides a good illustration of unified variety testing for regional adaptability. The goal is to provide a systematic flow of recommendations to farmers (Figure 15). Since 1977, most economic crops in the Philippines have been evaluated and introduced to farmers through this system. The system involves cooperation of the different research agencies, associations of colleges of agriculture, and the private sector.

Unified national testing of varieties and cultural practices is important for tomato improvement programs. Initially, variety trials might consist of elite breeding lines and accessions introduced from abroad. As more funds become available, locally bred lines and cultural practices can be evaluated regionally or nationally. Unified testing will foster cooperation among the country's tomato scientists and will facilitate exchange of germ plasm and information on common problems pertaining to the production and use of tomatoes. The trials would also serve as a mechanism for disseminating outstanding lines. Scientists

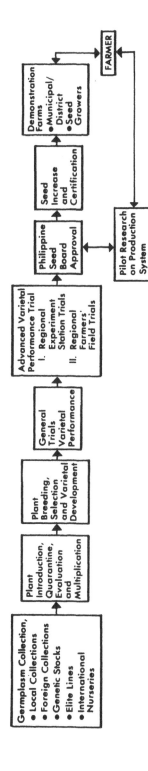

Figure 15. Organization of national crop seed improvement and extension in the Philippines. (Source: PCARR.)

involved in the trials should meet every year or so to discuss and summarize the results of previous trials, to eliminate poor performing lines, and to plan future trials with promising new lines. Such a group undertaking needs vigorous government support to maximize efficiency and benefits.

Appendixes

Appendix A:
Sources of Assistance

Aside from the Asian Vegetable Research and Development Center, which conducts tomato research and training, three international centers located in the tropics have research and training programs on intensive cropping systems in which tomatoes could be one component.

Asian Vegetable Research and Development Center (AVRDC)

The principal objective of AVRDC is to increase the yield potential and nutritional quality of crops that are particularly suited for supplementing the diets of Asian rice farmers and increasing incomes by diversifying farmers' year-round activities. AVRDC is located near Tainan in the coastal plain of southwestern Taiwan. The climate is tropical most of the year and vegetables are grown throughout the year, using modern agricultural technology.

AVRDC offers a wide range of research and production training opportunities to vegetable research workers and production specialists in tropical countries. The participants learn by doing at the side of the AVRDC's scientific staff. Upon return to their home countries, they strengthen national programs.

There are five categories of trainees at AVRDC.

- *Research interns.* Young men and women who have begun their careers in crop research and are currently involved with vegetable production in their home countries.
- *Research scholars.* Candidates for master's or Ph.D. degrees from colleges or universities that permit students to do their research at an international research center under the supervision of the center's staff scientists.

- *Research fellows.* Experienced scientists who generally have advanced degrees and who wish to familiarize themselves with new research techniques and information regarding vegetable improvement.
- *Production trainees.* A five-month course involves the training of vegetable production specialists who are responsible for training and supervising extension workers, farm managers, or educators.
- *Special trainees.* Special programs that are research or extension oriented, lasting a few weeks to a year, may be organized for certain individuals or groups.

The library maintains a full range of technical journals on vegetable technology and production, while the office of information services publishes an annual report and special reports. Technical bulletins and the *Guide* series written by scientists provide information about vegetable production technology for both research and extension workers. *About AVRDC*, a newsletter published three times each year, helps keep the international community aware of the center's activities. All AVRDC publications are free on request.

Asian Vegetable Research and Development Center
P.O. Box 42, Shanhua
Tainan 741, Taiwan

Centro Internacional de Agricultura Tropical (CIAT)

CIAT, located near Cali, Colombia, works primarily on cassava, dry beans, and beef cattle. Its small-farm systems program however focuses on family farms as integrated systems. Tomatoes are popular among Latin American farmers and are a good cash crop for small farms.

Training and other activities with national programs provide CIAT with feedback from cooperating countries on the functioning of farming systems that are relevant to the center's commodity programs. It accepts postdoctoral fellows and scholars on the master's or Ph.D. level to work with senior scientists.

CIAT publishes the results of its research, training, and international activities in an annual report. The farming systems section is available as a separate document. Extension publications are also available.

Centro Internacional de Agricultura Tropical
Apartado 67–13
Cali, Colombia

International Institute of Tropical Agriculture (IITA)

IITA, located near Ibadan, Nigeria, includes tomatoes in its farming systems program. Its major emphasis, however, is on improvement of grain legumes, root and tuber crops, and cereals for the humid and subhumid regions of Africa.

IITA offers research and production training for students, technicians, and scientists from many countries who work or who will work in tropical agriculture. Postdoctoral fellowships are offered to selected, qualified applicants with newly acquired Ph.D. degrees. Annual reports and other publications are available upon request.

International Institute of Tropical Agriculture
PMB 5320
Ibadan, Nigeria

International Rice Research Institute (IRRI)

IRRI, located in Los Baños, Philippines, does no work on tomatoes, but its cropping systems program, which concentrates on rainfed lowland and upland rice farms, offers possibilities for tomatoes to be grown by small farmers in the lowland tropics. IRRI sponsors a multiplecropping network in which different crop varieties are evaluated on rice-based farms in several locations throughout the Philippines and many parts of Asia. It offers a six-month multiplecropping training program and research training for young scientists, including postdoctoral fellows. Because of its proximity to the College of Agriculture of the University of the Philippines, qualified candidates may pursue an M.S. or a Ph.D. degree. IRRI publishes several reports, which are available upon request.

International Rice Research Institute
P.O. Box 933
Manila, Philippines

UNIVERSITIES

Scores of universities offer excellent graduate and training programs in vegetable crops, but only those with wide tropical experience or large tomato programs are mentioned here. Since most of these universities offer only a few assistantships or scholarships, candidates should have their own source of funding when applying for admission.

Cornell University

Half of the 450 faculty members of the College of Agriculture and Life Sciences and many others in related disciplines at Cornell have had substantial professional experience abroad. Many college faculty members devote a significant portion of their time to the international aspects of their disciplines. The Department of Vegetable Crops offers master's, Ph.D., and postdoctoral programs in the various aspects of vegetable production. It has trained and graduated several hundred U.S. and foreign students. The staff consists of physiologists, mineral nutritionists, management specialists, weed specialists, postharvest specialists, and breeders. Although New York State does not have a large tomato industry, the staff at the Cornell University Experiment Station and at the Geneva Experiment Station conduct many studies on tomato improvement and processing.

Department of Vegetable Crops
New York State College of Agriculture and Life Sciences
Cornell University
Ithaca, New York 14853, U.S.A.

North Carolina State University

North Carolina's Department of Horticultural Science has a strong faculty with tropical experience, particularly in Latin America. Graduate programs leading to the master's and Ph.D. degrees are offered with research specialization in various fields that may be oriented to tomatoes. Members of this department have been involved in collaborative research with the pathologists in developing tomato varieties with resistance to bacterial wilt. The state has a large fresh-market tomato industry. In addition to the research station at Raleigh, horticultural research is carried out at seven stations located throughout the state.

Department of Horticultural Science
North Carolina State University
Raleigh, North Carolina 27607, U.S.A.

Purdue University

Many faculty members in the Horticulture Department have substantial experience abroad, particularly in Latin America. The department also has a number of scientists who work on various aspects of the tomato (including ripening genes, nutritional quality, and breeding methodology), although Indiana does not have a large tomato industry.

Horticulture Department
Purdue University
Lafayette, Indiana 47907, U.S.A.

Texas A & M University

Horticultural research and extension efforts are centered at the main campus of Texas A & M at College Station. Horticulturists also are located at three large research and extension centers and several smaller research stations located throughout the state. Because of the great variations in soil types, climate, and vegetation of Texas and its proximity to Latin America, faculty members have experience with conditions similar to those in tropical countries. It has a program for developing tomato varieties for hot, humid conditions.

Department of Horticultural Science
Texas A & M University
College Station, Texas 77843, U.S.A.

University of California at Davis

The state of California has the largest contiguous tomato-growing area in the world. Basic and applied studies on processing tomatoes are being conducted with government and private (tomato producers and canners) support. The success of the state's industry is due largely to the continuous involvement of the staff of the Vegetable Crops Department and others at the Davis campus of the University of California. Many of its tomato scientists have had experience in tropical countries. The department conducts extensive studies on genetics, developmental breeding, practical breeding, and evaluation trials in many parts of California. The state has many seed companies and tomato canners where practical training also may be obtained.

Vegetable Crops Department
College of Agriculture and Environmental Sciences
University of California at Davis
Davis, California 95616, U.S.A.

University of Florida

The Vegetable Crops Department is a part of the University of Florida's Institute of Food and Agricultural Sciences. It is responsible for programs in undergraduate and graduate resident education in the College of Agriculture, research as a part of the Florida Agricultural Experiment Stations, service as a part of the Cooperative Extension Service, and international programs as a unit of the Center for Tropical Agriculture.

Because of the climate of Florida and the extensive and varied vegetable industry (tomatoes being the most important), the Vegetable Crops Department is highly capable of participating in programs of training and research in tropical and subtropical horticulture. Many faculty members located at the main campus at Gainesville and at the eight centers throughout the state have had experience abroad.

Vegetable Crops Department
University of Florida, IFAS
Gainesville, Florida 32611, U.S.A.

University of the Philippines at Los Baños

The College of Agriculture is the core of the University of the Philippines at Los Baños. The college has more than 200 faculty members who conduct research and instruct 2000 students (of these, more than 1000 are in the master's degree program and 100 are in Ph.D. studies). Several hundred students come from Asia, Africa, the Americas, and the Pacific. The Los Baños complex also includes a College of Forestry, the International Rice Research Institute (IRRI), the Southeast Asian Regional Center for Graduate Study and Research in Agriculture (SEARCA), and other institutions.

The college has become a center for training in tropical agriculture. Students from the tropics will find the conditions and training relevant to their local environments. Scholars of IRRI and SEARCA who are in degree programs obtain them at the university.

College of Agriculture
University of the Philippines at Los Baños
College, Laguna
Philippines

Appendix B: Some Experienced Tomato Scientists

This section lists some scientists who have had considerable experience with tomatoes in the tropics or who are involved with substantial tomato research programs. The author's familiarity with their work is another basis for inclusion.

Africa

Mr. C. S. Adam
Ministry of Agriculture and Land Use
P.O. Box 166
Mahe, Seychelles

Dr. B. O. Adelana
Institute of Agricultural Research
  and Training
Moor Plantation, Ibadan
Oyo State, Nigeria

Dr. Francis Agble
Crops Research Institute
P.O. Box 3785
Kumasi, Ghana

Mr. W. Baudoin
Centre for Horticulture Development
P.O. Box 154
Dakar, Senegal

Mr. A. Carpenter
FAO, Ministry of Agriculture
Zanzibar, Tanzania

Dr. W. J. Kaiser
East African Community
P.O. Box 30148
Nairobi, Kenya

Mr. J. G. Quinn
Ahmadu Bello University
Samaru, Zaria
Nigeria

Dr. S. Sinnadurai
Department of Crop Science
University of Ghana at Legon
Legon, Ghana

Dr. J. Uso
Crops and Soils Department
University of Nigeria
Nsukka, Nigeria

Mr. G. C. Wiles
Lowveld Experiment Station
P.O. Box 53, Big Bend
Swaziland

Dr. G. F. Wilson
International Institute of
    Tropical Agriculture
Oyo Road, PMB 5320
Ibadan, Nigeria

Asia

Dr. Juan C. Acosta
Philippine Packing Corporation
Camp Philips, Bukidnon
Philippines

Dr. Mahaboob Ali
Northern Mariana Agricultural
    Research Center
Kagman, Saipan, Mariana Islands
U.S. Territory in the Pacific

Mr. K. Blackburn
Plant Introduction and Horticulture
    Research Station
Laloki, Konedobu
Papua New Guinea

Dr. Mak Chai
Genetics and Cellular Biology
University of Malaya
Kuala Lumpur, Malaysia

Dr. B. Choudhury
Division of Vegetable Crops
    and Floriculture, IARI
New Delhi, 110012, India

Prof. Jose R. Deanon, Jr.
Institute of Plant Breeding
U. P. at Los Baños
College, Laguna, Philippines

Mr. Thaworn Govitayakorn
Department of Plant Science
Khon Kaen University
Khon Kaen, Thailand

Dr. K. M. Graham
Universiti Kebangsaan Malaysia
Jalan Pantai Baru
Kuala Lumpur, Malaysia

Prof. Han Huang
Department of Horticulture
National Taiwan University
Taipei, Taiwan

Dr. G. C. Kuo
AVRDC, P.O. Box 42 Shanhua
Tainan 741, Taiwan

Dr. C. R. Muthukrishnan
Faculty of Horticulture
Tamil Nadu Agricultural University
Coimbatore, 641003, India

Mr. Sujoko Sahat
Horticultural Research Institute
Passar Minggu, Jakarta
Indonesia

Dr. Hendro Sunarjono
Horticultural Research Institute
Passar Minggu, Jakarta
Indonesia

Dr. Thean Soo Tee
Malaysian Agricultural Research
    and Development Institute
Selangor, Malaysia

Dr. S. K. Tikoo
Indian Institute of Horticulture
    Research
255, Upper Palace Orchards
Bangalore, 560006, India

Dr. Thongchai Tonguthaisri
Maejo Field Crop Experiment
    Station
Sansai, Chiengmai
Thailand

Dr. Samson C. S. Tsou
AVRDC, P.O. Box 42 Shanhua
Tainan 741, Taiwan

Dr. Ruben L. Villareal
AVRDC, P.O. Box 42 Shanhua
Tainan 741, Taiwan

Dr. Manee Wivutvongvana
Department of Horticulture
Chiengmai University
Chiengmai, Thailand

Dr. Charles Y. Yang
AVRDC, P.O. Box 42 Shanhua
Tainan 741, Taiwan

Mr. C. H. Yu
Known You Seed Company
Kaohsiung, Taiwan

Dr. H. W. Young
Horticultural Research Institute
Lembang, Bandung
Indonesia

Latin America

Dr. Guillermo Hernandez Bravo
Instituto Nacional de Investigaciones
  Agrícolas, S.A.R.H.
Apartado 6–882 y 6–883
Mexico, D.F.

Dr. Flavio A. Araujo Couto
EMBRAPA
Caixa, Postal 1316
70.000 Brasilia D.F., Brazil

Dr. Cyro Paulino da Costa
University of Sao Paulo
Caixa Postal 83, 13.400 Piracicaba
Sao Paulo, Brazil

Dr. Julio Delgado
Boliche Experiment Station
Instituto Nacional de Investigaciones
  Agropecuarios
Guayaquil, Ecuador

Dr. Jose Fernandez
Universidad Central de Venezuela
Facultad de Agronomia
El Limon, Venezuela

Dr. Alejandro Ferrer Z.
University of Panama
P.O. Box 4714
Panama 5, Panama

Dr. Caroll E. Henry
Orange River Agricultural Station
Highgate, St. Mary
Jamaica

Dr. Miguel Holle
Centro Agronómico Tropical
  de Investigacion y Enseñanza
Turrialba, Costa Rica

Mr. Juan V. Jaramillo
Instituto Colombiano Agropecuario
Apartado Aereo No. 233
Palmira, Colombia

Dr. F. Kaan
Centre Reserche Agronomique
  des Antilles
Domaine Duclos
Petit Bourg, Guadeloupe

Dr. Jose A. Laborde
Instituto Nacional de Investigaciones
  Agrícolas, S.A.R.H.
Apartado Postal 112
Celaya, Gto., Mexico

Mr. Mario Lobo
Instituto Colombiano Agropecuario
Apartado Aereo No. 233
Palmira, Colombia

Mr. Fidel Lopez Lopez
Instituto Nacional de Investigaciones
  Agrícolas, S.A.R.H.
Apartado 356
Culiacan, Sinaloa, Mexico

Dr. Alfredo Montez
USAID/Florida Contract
U.S. Embassy
San Salvador, El Salvador

Dr. Hiroshi Nagai
Instituto Agronómico de Campinas
Caixa Postal 48, 13,000 Campinas
Sao Paulo, Brazil

United States of America

Dr. J. J. Augustine
University of Florida, IFAS
Agricultural Research and
  Education Center
Bradenton, Florida 33505

Dr. H. H. Bryan
University of Florida, IFAS
Agricultural Research and
  Education Center
Homestead, Florida 33031

Dr. J. C. Gilbert
Department of Horticulture
University of Hawaii
Honolulu, Hawaii 96822

Dr. H. R. Henderson
Department of Horticulture
North Carolina State University
Raleigh, North Carolina 27607

Dr. M. N. Jensen
Environmental Research Laboratory
University of Arizona
Tucson, Arizona 85721

Prof. P. W. Leeper
Texas A & M Experiment Station
Weslaco, Texas 78596

Dr. H. M. Munger
Department of Plant Breeding
  and Biometry
Cornell University
Ithaca, New York 14853

Dr. B. L. Pollack
Horticulture and Forestry
Rutgers, The State University
New Brunswick, New Jersey 08903

Dr. C. M. Rick
Vegetable Crops Department
University of California at Davis
Davis, California 95616

Dr. R. W. Robinson
Seed and Vegetable Sciences
  Department
Geneva Experiment Station
Geneva, New York 14456

Dr. W. L. Sims
Vegetable Crops Department
University of California at Davis
Davis, California 95616

Dr. W. H. Skrdla
ARS-SEA-USDA
Plant Introduction Greenhouse
Iowa State University
Ames, Iowa 50011

Dr. M. A. Stevens
Compbell Institute
P.O. Box 356
Davis, California 95616

Dr. E. C. Tigchelaar
Department of Horticulture
Purdue University
Lafayette, Indiana 47907

Dr. R. B. Volin
University of Florida, IFAS
Agricultural Research and
    Education Center
Homestead, Florida 33031

Glossary

Anthocyanineless. Absence of purplish pigment, as in the case of tomato seedlings with the *ah* gene.

Bacterial canker. A destructive disease caused by *Corynebacterium michiganense,* a bacterium that may be carried by the seed. Plants attacked by bacterial canker show many curled and withered leaflets, but the petioles remain attached to the stem. On the fruits, typical signs of the disease are bird's-eye spots with a halo that remains flat and white.

Bacterial spot. A disease caused by *Xanthomonas vesicatoria,* which enters the plant through the stomates or wounds made by wind, driven soil, insects, or cultural operations. It can be seed transmitted. Spot lesions can occur on leaflets, petioles, fruits, peduncles, calyx, and stems. Bacterial spot lesions on the fruits are raised and disappear as the spots become older.

Bacterial wilt. See *wilt.*

Breaker. A term to indicate change in tomato fruit color from green to tannish-yellow, pink, or red.

Cross-protection. An immune response to a virulent strain of a disease exhibited by a plant after initially being infected by a mild strain of the disease. The phenomenon has been more common among virus strains.

Early blight. A fungus disease of tomato caused by *Alternaria solani.* The fungus causes a stem canker or collar rot that damages young seedlings. Spots on the leaves appear as concentric rings. When the fungus attacks the fruits it may cause them to drop before they mature or they may develop dark, decayed spots as they ripen.

Emasculation. Removal of the anthers or male reproductive organ from a flower.

F_1. The first generation of a cross.

F_2. The second filial generation usually obtained by self-fertilization of F_1 individuals.

Hybrid. The product of a cross between unrelated varieties or breeding lines. F_1 tomato seeds can be planted only once since seeds obtained from such hybrid will not breed true and will result in yield reduction and non-uniformity.

Inflorescence. Floral axis consisting of several clusters.

Late blight. A disease caused by *Phytophthora infestans,* which also affects potato. The first signs of the disease are water-soaked or greenish blotches which appear on the leaves and other plant surfaces. The blotches enlarge rapidly, the plant withers and eventually dies. The disease attacks the stem, leaves and fruits of tomatoes.

Leaching. Movement of materials such as sodium, calcium, magnesium, sulfur, potassium, and nitrogen from the soil to the lower horizons due to excess water.

Locule. The seed cavity of a tomato fruit.

Lye. An alkaline solution such as sodium or potassium hydroxide. Used in the peeling of tomato fruits.

Mature green. The surface of the tomato fruit is completely green, but the fruit has reached the stage of development that will ensure the completion of the ripening process.

Mold count. The number of mold filaments that are present in tomato products (catsup, chile sauce, puree, paste, and juice). Mold count is used as standard for microorganisms in tomato products. A large number of mold filaments is unsatisfactory.

Mulching. The use of material such as straw, leaves, or plastics spread upon the ground to protect the roots of plants from excessive rain, to conserve moisture, to minimize weed growth, or to keep fruit clean.

Nightshade. A term commonly used to indicate plants belonging to the family *Solanacea*, such as tomato, eggplant, pepper, and white potato.

Oviposition. The act of laying eggs by insects.

Pathogen. A specific cause of a disease (bacterium, fungus, and virus).

Pedicel. The small stalk that holds the tomato fruit.

Pheromone. A substance produced by an animal (insects and mammals) that stimulates a behavioral response in other animals of the same species.

Pollination. The transfer of pollen grains from the male to the female reproductive organ of a flower.

Precursor. A limiting and necessary substance for the biosynthesis of another substance.

Systemic chemical. A chemical that is absorbed through the roots or through the leaves and is capable of being translocated from treated to untreated parts of a plant. It kills insects or may prevent infection by pathogenic organisms.

Thermophilic organism. An organism that is able to survive under high temperatures.

Virus. Submicroscopic infective agents that are capable of growth and multiplication in living cells and cause various diseases in tomatoes such as tobacco mosaic, cucumber mosaic, leaf roll, leaf curl, and spotted wilt.

Wilt. A disease in tomatoes that is usually caused by either bacteria, *Pseudomonas solanacearium,* or fungi, *Verticillium albo-atrum* and *Fusarium oxysporum.* The most common wilt in tropical countries is bacterial wilt, which is characterized by rapid wilting and drying of susceptible plants. A cross section of a tomato stem infected with the disease shows internal browning of the water-conducting tissue and the production of a glistening gray to yellowish bacterial ooze when placed in water.

Annotated Bibliography

Much of this book is based on references listed in this section, travel notes of the author, and conversations and correspondence with tomato scientists. All references are arranged alphabetically by chapters.

Chapter 1. Potential for the Tropics

Asian Vegetable Research and Development Center. 1976–1979. *Progress report.* 4 vols. Shanhua, Taiwan.
> Starting in 1976, AVRDC has published progress reports instead of annual reports. They summarize research, training, and outreach activities conducted by the center.

Asian Vegetable Research and Development Center. 1977. *Pre- and post-harvest vegetable technology in Asia.* Shanhua, Taiwan. 156 pp.
> Proceedings of a workshop held in the Philippines under the auspices of the Asian Vegetable Research and Development Center and Southeast Asian Regional Center for Graduate Study and Research in Agriculture. Major technical problems and the state of the vegetable industry in several Asian countries are described.

Asian Vegetable Research and Development Center. 1979. *Proceedings of the 1st international symposium on tropical tomato.* Shanhua, Taiwan. 290 pp.
> Contains 27 papers by authorities on tomato research and production. The articles cover such topics as tomato production in the tropics, improving small-scale tomato production, germ plasm resources, breeding for processing, fresh-market and heat-tolerant tomatoes, insect and disease control, and fertilizer application.

Cummings, Ralph W., Jr. 1976. *Food crops in the low-income countries: The state of present and expected agricultural research and technology.* Working papers. New York: Rockefeller Foundation. 103 pp.

> An analysis of the present state of research and knowledge of major food crops in low-income countries. Contains information on the activities of the international agricultural research centers with these crops. The author also identifies areas that need additional emphasis.

League for International Food Education. 1976. *Small-scale intensive food production.* Santa Barbara, Calif. 97 pp.

> A compilation of papers on the potential of small-scale intensive food production.

Munger, H. M. 1973. *Challenges of vegetable research for the tropics.* Mimeograph. Shanhua, Taiwan: Asian Vegetable Research and Development Center. 20 pp.

> Keynote address at the dedication of the Asian Vegetable Research and Development Center. Explores the importance of vegetable crops and the challenges that face vegetable scientists in tropical countries.

National Academy of Sciences. 1978. *Postharvest food losses in developing countries.* Washington, D.C. 206 pp.

> Summarizes information about losses of major crops and fish; discusses some of the economic and social factors involved, identifies major areas of needs, and suggests various policy and program options for developing countries and technical agencies.

Technical Advisory Committee Secretariat. 1976. *Report of the TAC vegetable research appraisal mission.* Rome: Food and Agricultural Organization. 25 pp.

> A brief review of tropical vegetable research in several countries of Asia and Africa from economic, nutritional, and social points of view.

Villareal, R. L. 1970. The vegetable industry's answer to the protein gap among low-salaried earners. *Sugar News* (Manila) 46:482–488.

> A discussion of the role of various vegetable crops for improving the Filipino diet. Also contains information on vegetable crops research at the University of the Philippines' College of Agriculture.

Chapter 2. Successful Programs

Asian Vegetable Research and Development Center. 1976–1977. *Tomato report.* 2 vols. Shanhua, Taiwan.

Summarizes the tomato research, training, and outreach activities of AVRDC.

Asian Vegetable Research and Development Center. 1979. *Proceedings of the 1st international symposium on tropical tomato.* Shanhua, Taiwan. 290 pp.
(See listing in Chapter 1 for annotation.)

Department of Agriculture and Forestry. 1975 and 1978. *Taiwan agricultural yearbook.* 2 vols. Taichung: Provincial Department of Taiwan.
Data on yield, area, production, and other important statistics on agricultural crops in Taiwan.

Grossman, B. D., and Korgan, G. 1977. *Agribusiness and economic aspects of Mexican vegetables moving to U.S. markets through Nogales, Arizona.* Department of Agricultural Economics, University of Arizona. 15 pp.
Reports the kind and volume of vegetables coming through Nogales, Arizona. Also contains information on the handling and destination of these vegetables.

Lai, S. H. 1975. Production of processing tomatoes in Taiwan. *Journal of the Chinese Society for Horticultural Science* 21, no. 2, pp. 1-8.
A brief description of the growth and development of the tomato processing industry in Taiwan.

National Food and Agricultural Council. 1975. *Implementing guidelines—Gulayan sa Kalusugan.* Quezon City, Philippines. 38 pp.
A description of the program to improve production, harvesting, collection, distribution, and marketing system of selected vegetables in the Philippines.

Oebker, N. F. 1977. South of the border. *American Vegetable Grower* 25, no. 9, pp. 17-18.
A brief report on the vegetable industry in western Mexico.

Olifernes, E. C. 1978. Coordinated tomato production and marketing in Northern Mindanao. *Philippine Journal of Crop Science* 3:235-237.
An account of a government-supported tomato production and marketing program in the southern Philippines.

Olifernes, E. C. 1978. *Financing the market operation of tomato in Cagayan de Oro City.* Mimeograph. Baguio City, Philippines: Bureau of Plant Industry, Region 10. 10 pp.

Pros and cons of various financing schemes for fresh-tomato production in the southern Philippines.

Rogers, H. T. 1979. Vegetable production—Mexican style. *American Vegetable Grower and Greenhouse Grower* 27, no. 4, pp. 48, 50–54, 56, 57; 27, no. 5, pp. 8–11, 14.
An account of winter vegetable production and marketing in western Mexico.

Taiwan Canners Association. 1973–1978. *Taiwan exports of canned food.* 6 vols. Taipei.
Data on the volume, dollar value, and destination of Taiwan's canned products.

Villareal, R. L. 1976. Observations on multiple cropping in Taiwan. *Philippine Journal of Crop Science* 1:129–136.
A discussion of the reasons for the success of multiple cropping in Taiwan.

Whitaker, T. W. 1977. Mexico's west coast winter vegetable industry. *HortScience* 12:535–538.
A description of the production and marketing of fresh vegetables from the west coast of Mexico, which supplies U.S. markets from October to June.

Yang, Y. K. 1977. *Farmers' organizations in Taiwan.* Taipei: Joint Commission on Rural Reconstruction. 26 pp.
A description of the farmers' associations and irrigation associations of Taiwan.

Chapter 3. Guidelines for Program Development

Asian Vegetable Research and Development Center. 1979. *Proceedings of the 1st international symposium on tropical tomato.* Shanhua, Taiwan. 290 pp.
(See listing in Chapter 1 for annotation.)

Benor, D., and Harrison, J. Q. 1977. *Agricultural extension. The training and visit system.* Washington, D.C.: World Bank. 55 pp.
Outlines a system of extending technology to farmers that is especially useful to developing countries.

Chandler, R. F., Jr. 1979. *Rice in the tropics: A guide to the development of national programs.* Boulder, Colorado: Westview Press. 256 pp.

Concise and plainly written to be easily understood by administrators and policymakers who plan and revamp agricultural programs of developing countries. Includes discussions of the more important elements of a successful rice production program and how these elements can be put in a workable scheme. The same elements could be useful in developing a general strategy for a tomato production program.

Villareal, R. L., and Wallace, D. H., eds. 1967. *Vegetable training manual.* College, Laguna, Philippines: University of the Philippines. 234 pp.

This manual is a handy reference for vegetable researchers and extension specialists in the tropics. Especially useful for tomato workers are chapters on planting vegetable crops, principles of fertilizer use, irrigation of vegetable crops, and seeds.

Chapter 4. A World Crop

Adams, C. F., and Richardson, M. 1977. *Nutritive value of foods.* Home and Garden Bulletin no. 72. Washington, D.C.: U.S. Government Printing Office. 40 pp.

Contains tables listing nutritive values for household measures of commonly used foods. More than 730 items analyzed for different nutrients.

Calkins, P. H. 1978. *Vegetable consumption patterns in five cities of Taiwan.* Technical Bulletin no. 5. Shanhua, Taiwan: Asian Vegetable Research and Development Center. 24 pp.

Essential reading for scientists planning to initiate studies on consumption patterns of tomatoes and other vegetables. The bulletin summarizes the food consumption patterns and trends in urban households and the role of vegetables in the diet.

Cruess, W. V. 1958. *Commercial fruit and vegetable products.* 4th ed. New York: McGraw-Hill. 884 pp.

An excellent book on the application of the fundamental sciences to the manufacture and preservation of fruits and vegetables.

Gould, Wilbur A. 1974. *Tomato production, processing and quality evaluation.* Westport, Connecticut: Avi Publishing. 445 pp.

Summarizes basic information on the production, processing, quality control, and evaluation of tomatoes and tomato products. An excellent source of information on processing and quality evaluation.

Goose, Peter A., and Binsted, Raymond. 1964. *Tomato paste, puree, juice and powder.* London: Food Trade Press. 151 pp.

A guide to the production of tomato products; describes various principles upon which the numerous processes are based.

Menegay, T. H. 1977. *The AVRDC vegetable preparation manual.* Shanhua, Taiwan: Asian Vegetable Research and Development Center. 105 pp.

Recipes and preparation tips for soybeans, mungbeans, sweet potatoes, chinese cabbage, and tomatoes. The manual should be useful for home economics agents, rural teachers, and public health workers throughout the tropics.

Rick, C. M. 1978. The tomato. *Scientific American* 239, no. 6, pp. 76–87.

Describes the important traits of the tomato, genetic manipulations, and new production technology that converted it from an exotic fruit into a popular food and a major article of commerce in the United States.

Urbino, Luisa M. C. 1973. *Household consumption patterns for fruits and vegetables, Philippines. 1970-1971.* Unpublished master's thesis, University of the Philippines. College, Laguna, Philippines. 100 pp.

Evaluates the consumption patterns of several vegetables in the Philippines. A useful guide for making similar studies in other developing countries.

Chapter 5. Varieties, Seed Production, and Distribution

Asian Vegetable Research and Development Center. 1974–1975. *Annual reports.* 2 vols. Shanhua, Taiwan.

Accounts of AVRDC's research and training activities on six crops and the international programs for 1972–1973 and 1974.

Asian Vegetable Research and Development Center. 1976–1977. *Tomato report.* 2 vols. Shanhua, Taiwan.

(See listing under Chapter 2 for annotation.)

Asian Vegetable Research and Development Center. 1976–1979. *Progress report.* 4 vols. Shanhua, Taiwan.

(See listing under Chapter 1 for annotation.)

Asian Vegetable Research and Development Center. 1977. *Pre- and post-harvest vegetable technology in Asia.* Shanhua, Taiwan. 156 pp.

(See listing under Chapter 1 for annotation.)

Asian Vegetable Research and Development Center. 1979. *Proceedings of the 1st international symposium on tropical tomato.* Shanhua, Taiwan. 290 pp.
(See listing under Chapter 1 for annotation.)

Campbell Soup Company. 1962. *Campbell plant science symposium.* Camden, New Jersey. 229 pp.
Papers dealing with the physiological and environmental factors affecting the reproductive process in plants; includes a thought-provoking analysis of fruit set in plants.

Castro, A. D., and Villareal, R. L. 1970. Effects of packeting materials on the moisture content, germination and vigor index of vegetable seeds in storage. *Philippine Agriculturist.* 54:241-276.
Shows the need to keep vegetable seeds in proper containers to maintain their germination and vigor under hot, humid conditions.

Graham, K. M.; Tan, H.; Chong, K. Y.; Yap, T. C.; and Vythilingam, S. 1977. Breeding tomatoes for lowlands of Malaysia. *Malaysian Applied Biology* 1:1-34.
An account of the development of bacterial-wilt resistant tomato breeding lines for the lowlands of Malaysia.

Herklots, G.A.C. 1972. *Vegetables in South-East Asia.* London: Allen and Unwin. 525 pp.
A comprehensive book on tropical vegetables, primarily directed to Southeast Asia, but of value throughout the tropics. Contains 159 drawings and hundreds of Chinese ideograms and vernacular names, which facilitate identification of many vegetables grown in the tropics.

Lobo, Mario A., and Jaramillo, J. V. 1977. Algunos aspectos sobre la producción de tomate. In *Curso Sobre Hortalizas.* Palmira: Instituto Colombiano Agropecuario. pp. 211-255.
A compilation of both basic and applied aspects of tomato production in Colombia.

Nishi, Sadao. 1966. "F_1 Seed Production in Japan." *Proceedings XVII International Horticultural Congress,* vol. 3, pp. 231-257.
The history, growth, and development of hybrid seed production of different vegetables in Japan.

Philippine Council for Agriculture and Resources Research. 1977. New system of crop variety testing for regional adaptability. *Monitor* 5, no. 8, pp. 1, 5-7.
  Recommendation for a national crop seed improvement production and extension scheme on farmers' field trials. A good model for developing countries that want to organize unified crop improvement and seed production.

Rick, C. M. 1978. The tomato. *Scientific American* 239, no. 6, pp. 76-87.
(See listing under Chapter 4 for annotation.)

Simons, J. H. 1978. Tomato improvement in South-Western Nigeria. *Vegetables for the Hot, Humid Tropics.* 3:37-41.
  Reports major research to improve tomato productivity in Nigeria.

Sunarjono, H.; Hartiningsih; Kirana, I.; and Sahat, S. 1976. Adaptasi varitas tomet untuk dataran rendah. *Bulletin Penelitian Hortikultura.* vol. 4, no. 4, pp. 3-11.
  A report of the performance of tomato breeding lines in the lowlands of Indonesia. Also discusses the problems of tomato production in that country.

Tee, T. S. 1978. *Vegetable production in Malaysia.* Selangor: Malaysian Agricultural Research and Development Institute. 21 pp.
  Discusses the problems of and government commitment to agricultural research programs for the improvement and development of the vegetable industry in Malaysia.

Tigchelaar, E. C. 1978. Tomato ripening mutants. *HortScience* 13:502.
  A description of three ripening genes in tomatoes that could have practical application in improving storage quality in tropical countries.

Tindall, H. D. 1970. *Prospects for horticultural development and research with particular reference to vegetable crops.* Addis Ababa: Institute of Agricultural Research. 44 pp.
  A report prepared by FAO on vegetable production in Ethiopia. Contains recommendations on staff development, research, and marketing of vegetables that should be useful in other African nations.

Tindall, H. D. 1977. Vegetable crops. In *Food crops of the lowland tropics.* eds. C.L.A. Leakey and J. B. Wills, pp. 102-105. Oxford: Oxford University Press.

A review of research and development for vegetable crops with emphasis in West Africa. Contains a section on tomatoes.

Villareal, R. L., and Wallace, D. H., eds. 1967. *Vegetable training manual.* College, Laguna, Philippines: University of the Philippines. 234 pp.
    (See listing under Chapter 3 for annotation.)

Villareal, R. L. 1971. In search of breakthroughs in vegetable production. In *Proceedings, in search of breakthroughs in agricultural development.* College, Laguna, Philippines: University of the Philippines, pp. 42-59.
    A discussion of progress in improving vegetable production in the Philippines; may be useful to other tropical countries.

Villareal, R. L. 1971. *Seed distribution systems.* Mimeograph. College, Laguna, Philippines: Society for the Advancement of the Vegetable Industry.
    A comparison of seed distribution systems in developed and developing countries.

Walter, J. M. 1967. Hereditary resistance to disease in tomato. *Annual Review of Phytopathology* 5:131-162.
    A review of diseases attacking tomatoes and progress in controlling them through breeding.

Yawalkar, K. S. 1965. *Vegetable crops of India.* 3rd ed. Nagpur, India: Agri-Horticultural Publishing House, 188 pp.
    Information on the major vegetable crops of India. One chapter summarizes the varieties, cultural practices, pests, and diseases of tomatoes.

Chapter 6. Cultural Practices

Asian Vegetable Research and Development Center. 1976-1977. *Tomato report.* 2 vols. Shanhua, Taiwan.
    (See listing under Chapter 2 for annotation.)

Asian Vegetable Research and Development Center. 1977. *Pre- and postharvest vegetable technology in Asia.* Shanhua, Taiwan. 155 pp.
    (See listing under Chapter 1 for annotation.)

Asian Vegetable Research and Development Center. 1979. *Proceedings of the 1st international symposium on tropical tomato.* Shanhua, Taiwan. 290 pp.
    (See listing under Chapter 1 for annotation.)

Bentley, M. 1973. *Hydroponics plus.* Sioux Falls, South Dakota: O'Connor Printers. 232 pp.

A simple book that describes growing of vegetables, herbs, or flowers in the living room, balcony, yard, or small farm through the use of hydroponics. It also includes full technical information for those who plan to go into commercial production.

Birch, M. C. 1974. *Pheromones.* New York: American Elsevier. 495 pp. Information on the biology of pheromones of insects and the mammals; concludes with a section on pheromones in manipulation of populations.

Chandler, R. F., Jr. 1973. The scientific basis for the increased yield capacity of rice and wheat, and its present and potential impact on food production in the developing countries. In *Food population and employment: The impact of the Green Revolution.* New York: Praeger Publishers. pp. 25–43.

A discussion of the scientific basis for increased yield capacity in rice and wheat. Outlines some practical innovations that can be initiated in developing countries to produce a significantly improved technology.

Dalrymple, Dana G. 1973. *A global review of greenhouse food production.* Foreign Agricultural Economic Report no. 89. Washington, D.C.: U.S. Department of Agriculture. 150 pp.

An introduction to the development, technology, and economics of controlled-environment agriculture in the form of greenhouse food production.

Dubois, P. (Translated and edited by C. A. Brighton.) 1978. *Plastics in agriculture.* London: Applied Science Publishers. 176 pp.

Summarizes important information on the development and use of plastics in agriculture and horticulture.

International Rice Research Institute. 1978. *Irrigation policy and management in Southeast Asia.* Los Baños, Philippines. 198 pp.

Seminar papers dealing with ways Southeast Asian countries are planning and designing irrigation infrastructure; management, operation, maintenance of irrigation systems, and training programs for water management personnel; economic issues in irrigation and irrigation organization and farmers' behavior.

Joseph, K. T.; Chew, W. Y.; and Tay, T. H. 1974. Potential of peat for agriculture. *Lapuran MARDI report* no. 16, pp. 1–16.
A discussion of peat soils of Malaysia and steps being taken by the government to make use of them for agriculture.

Knott, J. E., and Deanon, J. R., Jr., eds. 1967. *Vegetable production in Southeast Asia.* Quezon City: University of the Philippines Press. 366 pp.
Information on vegetable production in Southeast Asia with emphasis on the Philippines. In addition to a chapter on solanaceous vegetables, which include tomatoes, a discussion of principles and practices of vegetable production in the tropics is presented.

Lee, D. B. 1973. *Economic survey of fertilizer situation in the Asian and Pacific Region.* Taipei: Food and Fertilizer Technology Center. 198 pp.
An account of the fertilizer situation in Australia, Japan, South Korea, Malaysia, New Zealand, Philippines, Taiwan, Thailand, and Vietnam. Contains information on the production, trade, uses, and marketing of fertilizers in each of the countries.

Leon-Gallegos, H. M. 1978. *Enfermedades de cultivos en el Estado de Sinaloa.* Culiacan, Sinaloa: Centro de Investigaciones Agricolas del Pacifico Norte. 213 pp.
Lists the important diseases of major agricultural crops in the state of Sinaloa, Mexico. A chapter on tomato diseases summarizes the most destructive diseases and the recommended control measures.

Netscher, C. 1978. *Morphological and physiological variability of species of Meloidogyne in West Africa and implications for their control.* Wageningen: H. Veenman and Zonen. 46 pp.
Information on root-knot nematodes. Discusses the variability of *Meloidogyne* spp. in tropical soils of West Africa and recommends the use of nonhosts and resistant varieties as preventive treatments coupled with various chemical and physical treatments.

Sastry, K.S.M., and Singh, S. 1971. Effect of different insecticides on the control of whitefly (*Bemisia tabaci* Gen.) population in tomato crop and the incidence of the tomato leaf curl virus. *Indian Journal of Horticulture* 8:304–309.
A simple experiment demonstrating the economic importance of controlling tomato leaf curl virus through the use of systemic insecticides.

Sims, W. L.; Zobel, M. P.; and King, R. C. 1968. *Mechanized growing and harvesting of processing tomatoes.* AXT-232. Davis: Agricultural Extension Service, University of California. 28 pp.
A guide to mechanized growing and harvesting of processing tomatoes in California.

Stoner, A. K. 1971. *Commercial production of greenhouse tomatoes.* Agriculture Handbook no. 382. Washington, D.C.: U.S. Department of Agriculture. 32 pp.
Technical and practical considerations for a successful greenhouse operation. All data are from temperate growing conditions; nevertheless it should be useful for researchers and administrators in most developing countries.

Tee, T. S. 1978. *Vegetable production in Malaysia.* Selangor: Malaysian Agricultural Research and Development Institute. 21 pp.
(See listing under Chapter 5 for annotation.)

Tindall, H. D. 1970. *Prospects for horticultural development and research with particular reference to vegetable crops.* Addis Ababa: Institute of Agricultural Research. 44 pp.
(See listing under Chapter 5 for annotation.)

Tindall, H. D. 1977. Vegetable crops. In *Food crops of the lowland tropics.* Oxford University Press, pp. 102-105.
(See listing under Chapter 5 for annotation.)

Villareal, R. L., and Wallace, D. H., eds. 1977. *Vegetable training manual.* College, Laguna, Philippines: University of the Philippines. 234 pp.
(See listing under Chapter 3 for annotation.)

Villareal, R. L. 1971. In search of breakthroughs in vegetable production. In *Proceedings, in search of breakthroughs in agricultural development.* College, Laguna, Philippines: University of the Philippines. pp. 42-49.
(See listing under Chapter 5 for annotation.)

Walls, Ian G. 1972. *Tomato growing today.* Newton Abbot, Great Britain: David and Chales. 239 pp.
Discusses tomato cultural practices in the greenhouse in logical, systematic, and practical terms. Good source of information for tomato cultivation under controlled conditions.

Yawalkar, K. S. 1965. *Vegetable crops of India.* 3rd edition. Nagpur, India: Agri-Horticultural Publishing House. 188 pp.
(See listing under Chapter 5 for annotation.)

Chapter 7. Postharvest Technology and Marketing

Agricultural Marketing Service, 1973. *United States standards for grades of fresh tomatoes.* Washington, D.C.: U.S. Department of Agriculture. 1973. 10 pp.
Describes the different grade standards for tomatoes in the United States. A useful guide to policymakers in developing countries if grading of tomatoes is considered.

Darrah, L. B., and Tiongson, F. A. 1969. *Agricultural marketing in the Philippines.* College, Laguna, Philippines: University of the Philippines. 353 pp.
Reviews the marketing procedures of agricultural products, enumerates the problems involved, and presents the challenges and opportunities in a developing country. Contains several chapters on vegetables.

Grossman, B. D., and Korgan, G. 1977. *Agribusiness and economic aspects of Mexican vegetables moving to U.S. markets through Nogales, Arizona.* Tucson: Department of Agricultural Economics, University of Arizona. 15 pp.
(See listing under Chapter 2 for annotation.)

Mabilangan, Z. R. 1974. *Tomato marketing.* Quezon City, Philippines: National Food and Agriculture Council, Department of Agriculture. 28 pp.
A report on tomato marketing practices and channels in five regions of the Philippines. Concludes with the identification of problems and possible solutions.

Menegay, M. R. 1975. *Taiwan's specialized vegetable production areas: An integrated approach.* Technical Bulletin no. 1. Shanhua, Taiwan: Asian Vegetable Research and Development Center. 25 pp.
Describes the production and marketing of vegetables under the "Specialized vegetable production areas program" initiated by the Joint Commission on Rural Reconstruction. Explains the organization, structure, and implementation of the program.

National Academy of Sciences. 1978. *Postharvest food losses in developing countries.* Washington, D.C. 206 pp.
(See listing under Chapter 1 for annotation.)

Olifernes, E. C. 1978. Coordinated tomato production and marketing in Northern Mindanao. *Philippine Journal of Crop Science* 3:235-237.
(See listing under Chapter 2 for annotation.)

Pantastico, E. B. 1978. After harvest we also have to lower our losses. *Philippine Farmers' Journal* 20, no. 8, pp. 56-57, 60-63.
An analysis of the cause of losses in fruits and vegetables before and after harvest in the Philippines; relevant to other developing countries.

Rogers, H. T. 1979. Vegetable production—Mexican style. *American Vegetable Grower and Greenhouse Grower* 27, no. 4, pp. 48, 50, 52-54, 56 and 57; 27, no. 5, pp. 8-11, 14.
(See listing under Chapter 2 for annotation.)

Ryall, A. Lloyd, and Lipton, J. 1972. *Handling, transportation, and storage of fruits and vegetables (volume 1, vegetables and melons).* Westport, Connecticut: Avi Publishing. 473 pp.
Covers biological and physical aspects of marketing fresh vegetables.

Sims, W. L.; Zobel, M. P.; and King, R. C. 1968. *Mechanized growing and harvesting of processing tomatoes.* AXT-232. Davis: Agricultural Extension Service, University of California. 28 pp.
(See listing under Chapter 6 for annotation.)

Whitaker, T. W. 1977. Mexico's west coast winter vegetable industry. *HortScience* 12:535-538.
(See listing under Chapter 2 for annotation.)

Chapter 8. Promising Future Research

Asian Vegetable Research and Development Center. 1976–1977. *Tomato report.* 2 vols. Shanhua, Taiwan.
(See listing under Chapter 2 for annotation.)

Asian Vegetable Research and Development Center. 1977. *Pre- and postharvest vegetable technology in Asia.* Shanhua, Taiwan. 156 pp.
(See listing under Chapter 1 for annotation.)

Asian Vegetable Research and Development Center. 1979. *Proceedings of the 1st international symposium on tropical tomato.* Shanhua, Taiwan. 290 pp.
(See listing under Chapter 1 for annotation.)

Munger, H. M. 1973. *Challenges of vegetable research for the tropics.* Mimeograph. Shanhua, Taiwan: Asian Vegetable Research and Development Center. 20 pp.
(See listing under Chapter 1 for annotation.)

Philippine Council for Agriculture and Resources Research. 1977. New systems of crop variety testing for regional adaptability. *Monitor* 5, no. 8, pp. 1, 5–7.
(See listing under Chapter 5 for annotation.)

Technical Advisory Committee Secretariat. 1976. *Report of the TAC vegetable research appraisal mission.* Rome: FAO. 25 pp.
(See listing under Chapter 1 for annotation.)

Tigchelaar, E. C. 1978. Tomato ripening mutants. *HortScience* 13:502.
(See listing under Chapter 5 for annotation.)

Index

Pheromones, 128–129
Philippine Council for Agricultural and Resources Research, 17
Philippine Packing Corporation, 19, 23
Philippines, 4, 8, 11, 15, 16–23, 42, 55, 56, 64, 70, 74, 76, 78, 85, 89, 93, 96, 101, 111, 128
  Bureau of Plant Industry, 20, 90
  Claveria, 22–23, 43, 44, 50–51, 58, 97, 98
  Coordinated Tomato Production and Marketing Program, 18–23, 43, 46, 50–51
  Gulayan sa Kalusagan (GKP), 18, 19, 23
  Mamdahilin Farmers Cooperative, 18
  National Food and Agricultural Council, 17
  research and extension program, 17
  "Share for Progress Project" ("Green Revolution Program"), 17–18
  tomato yields in, 16, 22
Phytophthora infestans, 74, 124
Pickled tomato, 67
Plutella xylostella, 128
Pollination technique, 87
Portugal, 26
Postharvest handling, 114–115, 130
  hauling, 113
  packing, 35, 37, 112–113
  storage, 37, 113–114, 118–119
Postharvest losses, 11–12, 57–58, 111, 114
Prices, 11–12, 20, 22, 27, 29–30, 47–48
  establishment of, 44, 47
  seasonality of, 12, 16, 58
  See also Marketing of tomatoes
Private industry, 14–15
Problem soils, 42, 129
Processing tomatoes, 18, 24, 26, 29, 75

  harvesting of, 111–112
  international trade in, 7
Production
  areas, 56–58
  programs, 39, 40–51
  trends, 3, 15, 60
Protected cultivation, 58, 64, 93–96, 127
  influence of, on insects and disease, 93–94
  nylon nets for, 102
  for raising seedlings, 35–36, 98
  use of supports, 98
Pruning, 96, 98
Pseudomonas solanacearum. See Bacterial wilt
Pueraria phaseoloides, 105
Puerto Rico, 76
Pulp, tomato, 66
Purdue University, 140–141

Quality control. *See* Grades and standards
Quality of tomatoes, 75, 119
  breeding goals for, 79–80, 125–126
  for fresh market, 125–126
  individual preferences for, 69–71
  postharvest, influences on, 79, 114–115
  for processing, 65–67, 69, 75, 80, 126

Report of the Tomato Genetics Cooperative, 81
Research, 3, 13, 16–17, 38, 49–50, 65, 76, 122–133, 137–142
  funds for, 12, 16, 132
  importance of, 49–51, 121
  lack of, 12–13
  organization of, 132, 134
  relation to extension, 48–51
Resistance. *See* Breeding
Resources, analyzing availability of, 41–44